Virtues and Economics

Volume 6

The series is dedicated to virtue ethics and economics. Its purpose is to relocate economic theory to a domain where the connection between the virtues and economic decisions, as that connection is actually experienced in everyday life, is an organic component of theory rather than some sort of an optionally added ingredient. The goal is to help develop a virtue-based economic theory which connects virtues with the contents of economic activities of individuals, unincorporated and incorporated economic agents. The primary context is Catholic Social Teaching but other faith traditions (especially Judaism, Islam, Hinduism, Buddhism, and Confucianism) will also be explored for their construction of virtues in economic action. Special attention will be made to regulatory and policy issues in promoting economic justice.

The series connects virtue ethics with the core of economic theory and practice. It examines the basic and irreducible intentionality of human activities concerned with the production, distribution and consumption of goods and services. It considers the incommensurability of values as the central problem of economic decision making and examines whether that problem can be overcome by any means other than practical reason.

This series covers high quality edited volumes and monographs. The majority of the volumes are based on papers developed for the Economics as a Moral Science Program of Blackfriars Hall, University of Oxford.

More information about this series at http://www.springer.com/series/15627

Peter Róna • László Zsolnai
Agnieszka Wincewicz-Price
Editors

Words, Objects and Events in Economics

The Making of Economic Theory

Springer

Editors
Peter Róna
Blackfriars Hall
University of Oxford
Oxford, UK

László Zsolnai
Business Ethics Center
Corvinus University of Budapest
Budapest, Hungary

Agnieszka Wincewicz-Price
Polish Economic Institute
Warsaw, Poland

ISSN 2520-1794 ISSN 2520-1808 (electronic)
Virtues and Economics
ISBN 978-3-030-52672-6 ISBN 978-3-030-52673-3 (eBook)
https://doi.org/10.1007/978-3-030-52673-3

This Springer imprint is published by the registered company Springer Nature Switzerland AG.
The registered company address is: Gewerbestrasse 11, 6330 Cham, Switzerland.

Acknowledgement

The papers collected in this volume were presented at an international conference, titled Economics as a Moral Science, held in Jabłonna near Warsaw on 9 November 2018. It was organised and funded by the Polish Economic Institute.

The editors would like to express their gratitude to the Polish Economic Institute, the Mallinckrodt Foundation and Porticus for the financial support given to the publication of this volume as an open-source book, permitting free access to all. They would like to acknowledge the Polish Economic Institute's financial backing of the conference at which the papers included in this volume were presented.

Contents

Contributors

Ricardo F. Crespo is Professor of Philosophy of Economics at Universidad Austral and Universidad Nacional de Cuyo (Argentina) and Researcher at the National Scientific and Technical Research Council, Argentina. He earned a PhD in Philosophy and another in Economics (University of Amsterdam). Ricardo has extensively published articles and book chapters on his research topics. His last book is *Economics and Other Disciplines: Assessing New Economic Currents* (Routledge, London, 2017). His current research interests include explanation in the social sciences, economic rationality and ethics in economics.

Marcin Gorazda PhD, Lawyer, Philosopher, is Researcher at Copernicus Center for Interdisciplinary Studies, Jagiellonian University, Krakow, Poland. Marcin is a member of the Polish Philosophy of Economics Network. He specialises in philosophy of economics, philosophy of law, tax and commercial law. Marcin is the author of *Filozofia ekonomii* (2013) and *Granice naturalizacji prawa* (2017), both written in Polish.

Zsófia Hajnal is a self-proclaimed moral economist, a profession she pursues in parallel with her corporate career in international supply chains. She holds a BA degree in International Relations and an MSc in International Economics and Business from the Corvinus University of Budapest and the Budapest Business School, University of Applied Sciences, respectively. Ms Hajnal has also authored the paper 'Written and Unwritten Values in the Preamble of the United Nations Charter'.

Łukasz Hardt is an Associate Professor of Economics at the University of Warsaw, Poland, where he is the Chair of Political Economy at the Faculty of Economic Sciences. He is a member of the Monetary Policy Council at the National Bank of Poland. His latest book *Economics without Laws: Towards a New Philosophy of Economics* was published in 2017 by Palgrave Macmillan.

Brendan Hogan is Clinical Associate Professor in Global Liberal Studies at New York University and has spent 2018–19 as Research Fellow at Ecole des Haute Etudes en Sciences Sociale (EHESS). Broadly speaking, his research is focused on questions in the philosophy of social science, political philosophy, and pragmatism. Brendan's current project involves drawing out the consequences of pragmatism for understanding the relationship between social science and democracy. He has focused specifically on the model of rationality and action in economics and its impact on policy, institutions and social practices. His work has been published in the *Graduate Faculty Philosophy Journal*, *Human Studies*, *Journal of Speculative Philosophy*, *Pragmatism Today* and various volumes dedicated to issues in pragmatism and political philosophy.

Anna Horodecka is Assistant Professor in the Department of Economics (Collegium of Economic Analysis) at Warsaw School of Economics. She earned her PhD in the field of economic sciences (obtained in 2006 in Economics at the Warsaw School of Economics) and theology (2007). She specialises in economic policy, heterodox economics, methodology of economics and business ethics and is teaching these subjects in Polish, German, English and Italian. Anna received grants for research stays at Ludwig Maximilian University in Munich, Victoria University of Wellington in New Zealand and Northeastern Illinois University in Chicago and is a Visiting Professor at University of Macerata (Italy). She has authored and co-authored books, textbooks, monographs and articles published in Polish (e.g. National Economy) and international scientific journals. Her most important monographs are *The evolution of economic policy objectives* (2008) and *Concepts of Human Nature in contemporary economics* (2018).

Kevin Jackson JD, PhD, is Professor of Business Ethics and Law at Fordham University in New York City. Professor Jackson has served on the faculties of Georgetown University, Princeton University, Université libre de Bruxelles, École des Ponts and Peking University. Professor Jackson is a Senior Fellow at The Witherspoon Institute (Princeton) and a Fellow at European SPES Institute (Leuven). He has authored several books, including *Building Reputational Capital* (Oxford University Press), *Virtuosity in Business* (University of Pennsylvania Press) and *Charting Global Responsibilities: Legal Philosophy and Human Rights* (University Press of America) and regularly publishes interdisciplinary scholarship in philosophy, law, business ethics and musicology in top-tier academic journals. An acclaimed speaker for executives, dignitaries and academics around the world, Professor Jackson is also much loved by students for his interactive and high-energy presentation style. He composes and performs both classical and jazz music.

Maciej Kassner is a Political Scientist and Philosopher. He graduated from the Faculty of Political Science and International Relations at the University of Warsaw. Maciej is Adjunct at the Institute of Philosophy, University of Nicolaus Copernicus, Toruń. His research interests encompass political philosophy, philosophy of social sciences inspired by pragmatism, constructivism and hermeneutics, and institutionally oriented political economy. He is currently working on the interpretation of Karl Polanyi's political thought in the wider context of the contemporary controversies over the place of markets in modern society.

Eylül Seren Kösel is an independent economist working on research ethics. She earned a BSc degree in Economics from the Economics Department (Mülkiye) at Ankara University in 2018.

Tomasz Kwarciński is Assistant Professor in the Department of Philosophy, Cracow University of Economics, Krakow, Poland, Fellow of the Polish Economic Institute, Warsaw, Poland, Author of the first Polish monograph on Amartya Sen's theory of distributive justice (Równość i korzyść. Amartyi Kumar Sena koncepcja sprawiedliwości dystrybutywnej) and numerous articles in ethics, social philosophy and philosophy of economics (recently: A Hybrid Version of Well-being: an Attempt at Operationalization, "Public Governance", No 4 (46) (2018), 30–49).

Ivana Anton Mlinar (Mendoza, Argentina, 1977) is a PhD in Philosophy (Universidad Nacional de Cuyo -UNCuyo/Universität zu Köln) and a Researcher at CONICET (National Scientific and Technical Research Council, Argentina). She is currently teaching at the Instituto de Filosofía of the UNCuyo, does interdisciplinary research and publishes on Phenomenology, Philosophy of Mind and Cognitive Science.

Peter Róna obtained his degree in economic history from the University of Pennsylvania and his law degree from Oxford University. He was an Associate of the Washington D. C. law firm Arnold & Porter and Legal Counsel to the US Department of Commerce before becoming the Personal Assistant of Lord Richardson, Governor of the Bank of England. Peter joined the Schroder Group in 1969 as General Counsel of its operations in the USA and became the President and Chief Executive of The IBJ Bank & Trust Company in 1985. After the collapse of the Berlin Wall, he returned to his country of birth, Hungary, where he managed an investment fund. In 2003, he joined the Faculty of Eötvös Loránd University and was made an Honorary Professor in 2016. Peter joined the teaching staff of Blackfriars Hall in Oxford University in 2009. His published articles include a study of the euro and an examination of the philosophical foundations of economics.

Caspar Willem Safarlou is a Research Master's student in Philosophy at Utrecht University in the Netherlands. He specialises in ethics and its epistemological foundations. Currently, he is writing his Master's thesis on Philippa Foot's answer to the question 'why be moral?'

Liudmyla Vozna graduated from National Taras Shevchenko University of Kyiv in 1993. She received her PhD in Economics from the Institute of Economics at the National Academy of Sciences of Ukraine in 2001. Liudmyla was an Associate Professor at Zhytomyr State Technological University, where she taught economic theory, macroeconomics and regional economics. Liudmyla is an author of articles and book chapters devoted to the problems of economic freedom, the planned and transition economies, business cycles, economic stability and complexity, institutional economics, and regional economics. She is also the author of the book *The general Theory of the Socio-Economic Systems Stability* (2014) published in Russian, in which the central idea is the dispersion-probabilistic approach to an analysis of economic equilibrium.

Agnieszka Wincewicz-Price is Head of Behavioural Economics at the Polish Economic Institute in Warsaw. She studied at Warsaw School of Economics, Erasmus University Rotterdam (EIPE) and Newcastle University where she earned her PhD in Political Economy. Her research interests revolve around the intersection of heterodox and behavioural economics and philosophical anthropology, regarding questions of rational choice and decision theory, economic agency, variability and formation of preferences and their practical application in public policy. Agnieszka is Economics Editor of the peer-reviewed journal *Politics & Poetics*.

Altug Yalcintas PhD, is a philosopher of science and technology interested in the ethical economy of digitalisation and the Internet. He earned a PhD degree in Philosophy and Economics from the Erasmus Institute for Philosophy and Economics, Erasmus University Rotterdam, in 2009. Currently, he is a member of the Economics Department (Mülkiye) at Ankara University.

Wojciech Załuski is Professor in the Department of Philosophy of Law and Legal Ethics, Jagiellonian University, Krakow, Poland, and author of several books (recently: Law and Evil: The Evolutionary Perspective, Edward Elgar Publishing, 2018) and numerous articles in legal and moral philosophy.

Introduction

With all their graphs, models and puzzles that sometimes fill first-year students with awe and wonder, economics textbooks are relatively simple in their content. They sketch what seems to be a logical, if abstract, picture of the economic world, but they do not engage with any of the serious difficulties that in modern economic theory passes for economic practice and policy. You will not learn from them that economics does not have an ontologically objective subject, that economic life is the product of human (collective) intentionality or that there are more demanding elements in economic decision making than finding means to a given end.

The book before you, on the other hand, examines a long list of mysteries hidden from those elementary advances in economics that are reproduced more or less consciously by theorists, practitioners and students of economics who in their studies have not been daring enough to challenge or qualify received economic wisdom. Insofar, it is argued here, as these 'hidden' aspects of economic reality are important for its explanation and description, they should not be overlooked either by academic economists or by policy architects.

The papers collected in this volume address a wide spectrum of the apparent shortcomings of economic theory, generally falling into two related categories. One deals with the naturalistic pretence of economics that assumes the capacity to study economic realities in the same way as other disciplines study nature. This mistaken assumption results in other misleading postulates, the most harmful of which is probably that the subject-matter of economics can be treated as ontologically independent both of human intentions and from human purposes. The other category of shortcomings deals with the consequences of the attempt to render economics a value-free and predictive science. This allows paradoxes to emerge, for example, the inability of economic theory to account for the central concepts of economics, such as well-being, in meaningful and normative terms. The values of daily life associated with well-being, such as *bona fide* dealings and natural justice (e.g. the prejudice in favour of promise-keeping), are excluded because they cannot be quantified and thereby measured or mathematically expressed.

The trouble, however, is that the language employed to describe the economic world – as well as judgements of relevance necessary for its analysis – are value laden. When we think of welfare, it is impossible to strip it of all its normative content without seriously altering its meaning. Assuming that a person's welfare is self-centred, for example, it depends only on her consumption bundle, and that the only goal of a person is to maximise her welfare (as well excluding the possibility that a person's welfare can depend other people's well-being) is to make strong normative statements which often fly in the face of facts about people's motivations and desires. This is just one example of a much broader problem of the entanglement of descriptive and normative elements in the language and content of economics.

This challenge is often overcome, or at least concealed, either by the formalisation of economic concepts in mathematical terms or the rendering of the subject matter in the form of mathematical theorems. The ubiquitous use of mathematical formalism in economic analysis is a consequence of its assumed naturalistic orientation and of the attendant modelling of its methodologies on those of various natural sciences. Methodological borrowing is justified by the apparent resemblance of economic and natural phenomena: both are said to be subject to laws and regularities that are ontologically independent from human will. But this putative resemblance rests on the restatement of the subject matter in terms that exclude any and all human-centred values. The neoclassical school of economics, which, despite its commonly recognised flaws, remains the core of the economics curriculum, teaches that it is possible to explain economic phenomena using theoretical models that isolate the economic motives of actors from other factors conditioning the socio-economic world. These models treat the latter as unnecessary, 'disturbing causes'. The truth-value of these constructed (read: 'non-existent') objects of economic inquiry becomes a matter of their consistency with hypothetical assumptions rather than empirical verification.

In opposition to that approach, one of the major themes of this book is the ontological subjectivity of economics whose reality is constituted by institutions. Institutions in turn are collectively, that is to say, socially, created by means of language, in shared or intersubjective meanings and interpretations. Economic objects are here shown to be innate in human intentionality and judgement; they emerge from human activity with moral purposes embodied in them. Economic phenomena are traditionally interpreted as the instantiations of laws. However, they turn out merely to be statements describing the behaviour of theoretical economic models. The use of models to explain the workings of economic systems is shown inherently to be flawed.

The four preceding volumes of this series have covered the limitations of economic models and consequences thereof. Several essays in this collection focus on one particularly important weakness of economic models: their limiting assumptions regarding the identity and reasoning powers of economic agents, particularly the human agent, has come to be defined as no more than a collection of preferences. As the subject of economic science and analysis, human beings have lost their distinctiveness as persons, as autonomous moral agents, having been replaced by impersonal economic agents, without a will other than the one supplied by economic theory, and therefore necessarily amoral. This too is a result of the naturalistic orientation that denies economic agents the capacity of practical reason. In consequence, it also denies them the need and ability to learn what is good for them and to those around them, and to make value choices in accordance with the sort of person they consider themselves to be.

Typical dismissals of the problems posed in the forthcoming essays are rhetorical questions like: Why should economists bother? Why not leave the unending criticism to philosophers, sociologists, and perhaps theologians? They like to busy themselves with theoretical quandaries without seeming to understand the practicalities of the economic world. A familiar answer is the well-known fact that where theory meets the real world, economic models turn out to be unsuitable or the insufficient source of knowledge about the causes, nature and consequences of the observed economic phenomena. It becomes evident, for example, that economic actors facing situations of choice cannot be sure what the consequences of alternative lines of action will be. With imperfect information or incomplete markets, rational economic calculations might be impossible to perform, and instrumental rationality, such as means-ends schema, could not be relied on. Since unquantifiable uncertainty is one of the most enduring features of our contemporary economic life, persons do not routinely act rationally, they rather base their decisions on their beliefs about what they trust to be right. To further complicate things, much of their decision making is streamlined by institutions which form a specific embodiment of collective intentionality and collective rationality. Theories and models that do not account for these complexities are bound to be contradicted by empirical data. If used as basis for policy prescriptions, they might and indeed do become sources of harmful practices – think of recent debates about austerity measures.

In recent years, however, an even more serious objection to modern economic theory has come to the forefront. The paradigm theory contrived for authorised economic behaviour, namely the utility-maximising economic agent, is now seen not only as mistaken but as downright destructive. This problem was recognized already by the initiators of what came to be known as Corporate Social Responsibility. The utility-maximising agent has, understandably, acquired a bad reputation, but there is also a growing perception that we need to reject the myth of universal prosperity being arrived at chiefly by individuals pursuing self-chosen good.

The upshot of the reflections undertaken in this volume is that accurate understanding of the claims made by economics and of the methods employed by it requires a deep engagement with its underlying theoretical and philosophical presuppositions. Otherwise, economics is bound to stray further not only from practical applicability to real-world phenomena but also from our essential humanity.

As a start, it might help to conceive of economics as a discipline based on capacities, tendencies and undetermined potentialities rather than strict laws of nature. Economic explanation would also gain from a broader and therefore better understanding of the nature of human choice, its dispositions to virtue and vice, and its motivational complexity. What follows from this is the need to approach economic modelling in ways that do not abstract from essential aspects of human nature, particularly because they do not lose the ability to identify the economic agents under investigation as a human person. If it is claimed that a model has a realistic connection with the world it describes, then we must be able to establish 'to what' and 'to whom' the description applies.

Lastly, it is rarely realised but should always be emphasised that economics as a scientific discipline joins other social and human sciences in their search, or rather their specification, of what forms the ultimate end of social endeavours and policy prescriptions. In economics, the ultimate end falls under the rubric of the concept of well-being, whose content is undecidable on any simple, abstract or deterministic model. Its specification requires an understanding of the culturally distinct and socially constructed character of human needs, desires and preferences as well as enlisting the values and moral perspectives of economic subjects themselves. Such an approach surpasses the limits of scientific logic and blends into the logic of normative – that is, ethical and moral – justification. It employs an axiologically richer account of human motivation, which goes beyond the calculation of individual, future-oriented gains; it exceeds the limits of utility maximisation, allowing space for people's need and desire to do the right thing. It is important to realise that most words, objects and events in economics are ultimately embedded in this permanent endeavour.

Oxford, UK Peter Róna
Warsaw, Poland Agnieszka Wincewicz-Price

Chapter 1
Made with Words

Peter Róna

"And wonder if Man's consciousness
Was a mistake of God's."

(Thomas Hardy, *I Travel as a Phantom Now*)

Abstract The natural sciences study objects; the social sciences are concerned with events. Objects have properties, events are the products of human intentionality. Modern economics is built on the assumption that events can be transformed by a process of abstraction into things with properties of the sort that permits the formulation of law-like generalities. These assumptions have resulted in the paradigm of the value-free economic agent. This paper rejects this paradigm and defends the claim that the domestication of the methodology of the natural sciences in economics is a category mistake. Instead, economics should develop a methodology that is proper to its subject.

1.1 Introduction

The ambition of modern economics to resemble to the maximum possible extent the natural sciences involves the reconstruction of economic processes into objects that resemble the objects of the natural sciences. This move permits the use of causation theories that are analogous to causation in the natural sciences where causation depends on the properties of the objects participating in the phenomenon under study. But economic events, unlike the objects of nature, come about through human effort and intention. Their characteristics are given not by nature but by the human agent. Natural objects have no meaning; human intentions are inherently value

P. Róna (✉)
Blackfriars Hall, University of Oxford, Oxford, UK
e-mail: peter.rona@bfriars.ox.ac.uk

P. Róna et al. (eds.), *Words, Objects and Events in Economics*, Virtues and Economics 6, https://doi.org/10.1007/978-3-030-52673-3_1

laden, the product of human will and intellect. Modern economics excises the human will and intellect from the domain it studies and treats the remainder as if it were the sort of object one finds in nature, such as molecules, compounds, trees or the electrical current. With the indispensable help of this excision, modern economics formulates law-like generalisations that portray economic events as phenomena without moral content. But this outcome is fundamentally at odds with human experience, and it is certainly not supported by either its predictive or explanatory record. The result is that the actual processes of economic life, resting as they do on human will and intellect regularly defeat the predictive and explanatory power of modern economics, constructed to be innocent of that will and intellect. The tension is obvious. Either there are economic laws of universal applicability or economic life is, at least in part, the life of the human mind and will. The third alternative, namely that all humanity is of the same mind and will, does not command the assent anyone who is not a professional economist.

The title of this paper, "Made with Words" comes from Thomas Hobbes and from an era very much alive to the ontological difference between the natural and the moral sciences where the former is concerned with the properties of corporeal objects as opposed to the behaviour of incorporeal, or moral entities.[1] The XVIIth and the first two thirds of the XVIIIth century, still under the influence of Cartesian dualism, firmly believed in the existence of both a physical and a non-physical world, but by the end of the XVIIIth century the unity of nature had become axiomatic, and the idea that the scientific method of the natural sciences could be extended to social reality was already propounded by Newton and heartily endorsed by, among others, Diderot and D'Alembert. As Newton put it, "if natural philosophy in all its parts, by pursuing this Method, shall at length be perfected, the bounds of moral philosophy will be also enlarged." Throughout the XIXth and the XXth century, economics, born with Adam Smith and the dawn of Rationalism, turned to the notion of the unity of nature and the applicability of the methods of natural science to all of reality with increasingly ungovernable fervour. Einstein claimed that economics could be reduced to microphysics, and some Logical Positivists thought that sociology was no more than a particular instantiation of molecular genetics. The unity of nature axiom spelled for much of the XIXth and the XXth centuries the eclipse of ontology and, correspondingly, introduced a long period during which the *nature* of the subject matter under study, the examination of its properties was largely ignored. The eminent Harvard philosopher, Hilary Putnam went so far as to write as recently as 2004 that ontology was "a stinking corpse" in need of urgent interment.

But surely, if Galileo Galilei was right in claiming, as he did, in his *The Assayer,* that the book of nature is written in the language of mathematics there are other books about other realities and about human experience that are not written in that language. (That he was not right was shown brilliantly by Eugene Wigner in 1959,

[1] Hobbes' phrase was extensively explored in the context of political theory by Pettit (2008).

but that is another story.)[2] So, what is the nature of economic reality and how does it differ from natural objects?

1.2 The Objects of Nature and the Objects of Thought

The objects of nature exist independently of any theory or hypothesis about them. Their properties are unaffected by any human thought or action. It is this characteristic of natural objects - their independence from and indifference to any theory about them - that makes it possible to formulate verifiable theories about them. Substances exist regardless of what we make of them, and their transformation under various circumstances is the consequence of their properties interacting with the properties of other natural objects rather than the human attribution of such properties. Their nature cannot be changed by regulation or persuasion. They are impervious to incentives. It is a fundamental requirement for the validity of any scientific theory that the object of the theory be separate from the theory about it. Scientific experiments are repeatable precisely because the properties of the subject matter of the experiment are constant, and, moreover, they are universal. The valences of the elements in Mendeleev's table are the same in Denmark, Argentina or the Philippines.

In sharp contrast, the objects of economic activity are objects of thought. They are mind-, or as Russell has it, representation dependent,[3] brought into being through collective intentionality. Although the objects of thought may result in the production, distribution and consumption of tangible, corporeal objects, the object of economic thought itself is incorporeal. A 'market' may be a physical space, but the economically significant aspect of a market is not its physical configuration, not its physical properties, but, rather, the properties of the activities carried on therein. The economically significant properties of physical objects are not their physical properties, but the terms and conditions for their ownership, exchange and the conditions for the determination of their value. These intangibles are *not* properties of the objects themselves, but, rather, they are the generally accepted social conditions for their creation, ownership, use and consumption. Economic objects come into being as the fruit of human thought relating to both their conception as objects as well as the social organisation that permits their creation and sustenance. These objects are social facts, the embodiments of collective intentionality, emerging from a collective ethos and they come into being through language, reflecting the thought that gives rise to them. Because they are the product of human intention, they are inherently subjective, reflecting the intent of those who create them. Thus, 'money', a piece of paper with specific size, colour design and printing is endowed with specific properties and functions by the collective intentionality with which it is consti-

[2] Wigner (1960, 1–14).

[3] Russell (1919)

tuted and with which it is continuously redefined. The piece of paper is an objective or corporeal object, but its significance and role are not a function of its physical being. Indeed, money no longer needs to have any physical form at all.

1.3 What Is Intentionality?

Intention, including intentional action, is one of the most complex aggregate of issues in philosophy, involving the mind/body problem, causation, knowledge, truth, practical reasoning, ethics and ontology. Economic events, the complex aggregate product of intentionality – including opposing or contradictory intentions – are the products of intentional human action.

Philosophical interest in intentionality, developed first by Aristotle's medieval followers, most notably St. Thomas Aquinas, was largely forgotten until revived by Brentano, Husserl and Meinong. The publication of G. E. M. Anscombe's *Intention*[4] combined the Aristotelian – Thomist tradition with analytic philosophy in place of Brentano's empiricism. Since then the philosophy of intentionality has been greatly expanded by, among others, Peter Geach,[5] John Searle,[6] Graham Priest[7] and Tim Crane.[8]

A review of the many theories and subtleties of intentionality is beyond the scope of this paper, and for our purposes the simplest definition of the notion will suffice. It is the capacity of the mind to direct itself onto any object, be it existing or not existing and be it a concept or a thing. For the narrower purposes of this paper intentionality is the name of the mental constitution and representation of the habits, beliefs and rules with which objects are endowed with exchangeable value, how that value is created, preserved, diminished and exchanged, in short Adam Smith's famous human „propensity to truck, barter and exchange".

Whereas modern economists, such as the Nobel Laureate N. Gregory Mankiw believes that economists "approach the subject of the economy in much the same way as a physicist approaches the study of matter and a biologist approaches the study of life... (economists) devise theories, collect data, and analyse these data in an attempt to verify their theories",[9] Smith sees the heart of the matter in the „offering of an argument to persuade one to do so and so."[10] These two conceptions of economics could not be further apart. According to Mankiw, the subject matter is much like matter, harbouring data which is there to be collected for the verification

[4] Anscombe (1957)

[5] Geach (1967, 627–32)

[6] Searle (1983)

[7] Priest (2005)

[8] Crane (2013)

[9] Mankiw (2011)

[10] Smith (1978)

of some causal generalisation. For Adam Smith it is about the argument to persuade, given our propensity to truck, barter and exchange. Mankiw believes that economic objects have an objective, independent existence. Smith, in contrast, believes that they exist only by virtue of human propensities, that the economy is not a collection of objects but an array of human propensities.

A further specification is that the intentionality at play in the economy is collective. The intentions of the individual agent operate in the context of a collective, socially constructed and socially legitimated intentionality. There are no economic events within the domain of a single person. Whereas the protagonist of modern economics is the utility maximising *homo oeconomicus* with a single motive for his actions and the exclusive prerogative to determine his utilities, in fact, utilities are social constructs and so are the rules with which choice among available utilities may be made.

The intentional objects of human economic activity, Smith's propensities, are incorporeal, rendering the usual techniques of abstraction and idealisation regularly employed in the natural sciences problematic. The elimination of „disturbing causes" – to use Mill's famous phrase, – leaves after the elimination some corporeally identifiable object with verifiable properties. But in the case of incorporeal objects this is not possible. The conceptualisation of experience may produce a useful metaphor or icon for the sum of that experience, but it is not its distillation through the elimination of eliminable components because such components are themselves abstractions. A demand curve is an effective visual representation of the abstraction called demand, but it is not the sum and cannot be the sum of empirical data. As Nicholas Kaldor noted, the enemy that defeats the successful summary is time. Demand for a good takes shape over time, but during the same period other relevant factors, such as the availability of substitute goods, a change in interest rates or the level of confidence, the operation of monetary policy, etc. alters the context in which the demand takes place, and the demand curve cannot account for such changes. Economists deal with this problem with the help of the *ceteris paribus* assumption, but the conceit that everything else remains constant is just that, an arbitrary conceit. The intuition that there is a relationship of some sort between prices and quantities is undoubtedly a useful generalisation, but it does not have the sort of truth value that a typical law in physics or any branch of the natural sciences carries.

1.4 Is This Just a Question of Complexity?

It is frequently claimed that the difference between the facts of nature and social facts is one of quantity rather than quality. The facts leading to price formation, according to this view, are qualitatively the same as those that produce nuclear fission. The difference is only that there are more of them. Social facts involve greater complexities because they are the composites of a much larger number of components than the facts of nature. But this claim is mistaken for two rather fundamental reasons. First, social facts are intentional facts, depending on human intentionality, whereas natural facts do not contain any intentionality. Second, social facts operate in an open world where new facts come and old facts go without there being any principle regulating the coming and going, in contrast with natural facts where the context, as in a laboratory experiment, can be determined. The difference between facts that come into being through human agreement and custom and facts that are totally independent of us is quite basic to science. It is the precondition of the separation of the experimentally verifiable theory from its object. If monetary theory were about the properties of the piece of paper serving as a ten pound note, it would have no difficulty in resembling the natural sciences. But it is not about that piece of paper. Monetary theory aims to be about the *functions, purposes, effects and responses in individual and collective behaviour* we collectively ascribe to the ten pound note together with the rules, the terms and conditions we explicitly and implicitly posit that renders that note operational. *That ceaseless variation in the value of that ten pound note is the consequence of the ontological subjectivity of the collective intentional object we call money.* As Adam Smith had it, the value of that ten pound note is a matter of persuasion, a matter of how, through that dialogue between buyer and seller and the ceaseless interaction within society at large value is arrived at, how they reach agreement that the note is worth more or less or just about the same as the object for which it is exchanged.

1.5 Do Economic Objects Exist?

One of the most intensely debated issues in the philosophy of intentionality is whether the object of thought must, in fact exist in order to be the object of intentionality. The materialist view answers the question in the affirmative, claiming that the existence of imagined objects is an altogether different sort of existence from real existence. Others, like Tim Crane, hold that, inasmuch as nonexistent objects can have specific properties, they can be the objects of intentionality. If it is reality that grounds the truth of a claim and if reality does not contain more than what exists, the question what objects exist becomes quite important. If economic objects exist in the same sense as molecules, the determinism of modern economics is well founded. Economists can then claim to be making discoveries about the world much like any natural scientist. Buti if economic objects do not exist, there is nothing that can be discovered about them and the claim to science collapses. Non-existent objects have no laws of nature and cannot have causal powers because causal powers require spacio-temporal locations, something non-existent objects do not have.

The non-existence of economic objects – their lack of ontological objectivity – does not, however, preclude the possibility of constructing objects with a subjective ontology. Sherlock Holmes, Pegasus and the planet Vulcan are all non-existent objects, and yet, it is possible to make true as well as false statements about them. It is, for example, true that Sherlock Holmes was a detective, and it is not true that he was an aerobics instructor. It is true that Pegasus has two wings but not true that it won the Kentucky Derby, etc. In the same way, demand curves, 'markets' or 'marginal propensities are no easier to locate in the real world than Sherlock Holmes, but it is quite possible to make true as well as false statements about them. The demand curve is a line in a textbook the shape of which is not the property of any object, but, rather, the visual rendering of a collection of generalised characteristics of the relationship between prices and quantities. A market may be a place – and therefore a real object – but need not be in order to possess characteristics of the processes taking place in real markets. The problem facing economics is compounded by the fact that, whereas the three non-existent objects mentioned above are corporeal, this form of fictional corporeality in typically unavailable to economic objects. That corporeality becomes important when the truth claims about Sherlock Holmes may have truth value if they are consistent with the postulated axioms about him. A claim that he can fly will not satisfy this condition. Pegasus on the other hand may well be able to fly but it cannot smoke a pipe, converse with Watson or solve a crime. The question for economics therefore, is how to determine the truth value of claims about non-existent intentional objects where the object would be ontologically objective if it existed, but where the object not only does not exist, but does not have even fictional corporeality. The short answer is that it cannot be done in compliance with the requirements for truth value in the natural sciences because *truth value in the natural sciences is a function of experimental confirmation while the truth value of non-existent objects is a matter of consistency with assumptions and postulates.* (A concrete illustration will be given below in the discussion of Walras.)

Uncertainty about the ontological objectivity of the entity under study leads to a failure in demarcating the distinction between theory and its subject matter. This failure deprives the theory of its scientific status and leads to the sort of circularity noted by Joan Robinson: „*Utility* is a metaphysical concept of impregnable circularity; *utility* is the quality in commodities that makes individuals want to buy them, but the fact that individuals want to buy them shows that they have *utility*."[11] A comparison of the definition of chemistry set out by Robert Findlay Henry with the two on offer from Adam Smith and Mankiw quoted above will reinforce this point. According to Henry, „Chemistry attempts to understand transformations between substances."[12] If we follow Adam Smith in attempting to formulate a similar definition of economics, and claim that „economics attempts to understand the nature and causes of the wealth of nations", the question arises whether 'substances' are the same as 'wealth', and whether 'transformations' pose an analogous challenge to Smith's 'nature and causes' and whether the sort of understanding posited by Hendry is the sort of understanding that can be had of wealth.

Substances exist regardless of what we make of them, what utility we apprehend in them, and their transformation under various circumstances takes place quite independently of human action. But wealth, utility, equilibrium, inflation and so on

[11] Robinson (1962)

[12] Hendry (2008)

are socially constructed – they are 'social facts'[13] – subject to significant and more or less continuous reconstitution. Social facts cannot be separated from the theory or rules under which they are formulated.

The difficulty of separating subject from theory where the subject has no independent ontology is also apparent in Lionel Robbins' famous definition: „Economics is the science which studies human behaviour as a relationship between ends and scarce means which have alternative uses."[14] But 'ends' and 'scarce means' are socially constructed non-existent objects and the 'human behaviour' under study is an integral part of that social construction. It is human behaviour that constructs both the end and the scarcity, and it is the ends and the scarcity that constructs human behaviour. We are back to Joan Robinson's „impregnable circularity".

Physical objects, such as the chemist's elements do not have a purpose, an end, a *telos*. The purpose with which they may be endowed, the use to which they are put is not a property of these elements. Aristotle's famous unsolved problem of the incommensurability of use value and exchange value – as indeed the constant problem of incommensurability in economics – cannot be solved if we are searching for a 'value' that is common to both.[15] His beds and shoes do not have either use or exchange values. Such values are attributed to these objects by the intentionality of the exchanger or user. Whereas the valence of a chemical element is a *property* of that element, exchange and use values are attributions depending upon the intention of the agent. The value is in the intentionality as opposed to the valence which is in the object. Keynes was acutely aware of the problem but did not quite know what to do about it. In a letter to Roy Harrod he wrote:

> In chemistry and physics and other natural sciences the object of experiment is to fill in the actual values of the various quantities and factors appearing in an equation or a formula; and the work when done is once and for all. In economics that is not the case, and to convert a model into a quantitative formula is to destroy its usefulness as an instrument of thought … I also want to emphasise strongly the point about economics being a moral science. I mentioned before that it deals with introspection and with values. I might have added that it deals with motives, expectations, psychological uncertainties. One has to be constantly on guard against treating the material as constant and homogeneous, even as the material of the other sciences, despite its complexity, is constant and homogeneous. It is as though the fall of the apple to the ground depended on the apple's motives, on whether it is worth its while falling to the ground, and whether the ground wanted the apple to fall.[16]

Although Keynes comes very close to recognising intentionality as the ontologically decisive difference between the moral and the natural sciences, he does not solve the dilemma implicit in modern economics[17]: are economic objects like

[13] Gilbert (1989)

[14] Robbins (1932). In a later version the puzzling phrase „behaviour as a relationship" was replaced with the notion of „allocation".

[15] Meikle (1955)

[16] Letter to Roy Harrod, of 10 July 1938 (misdated 16 July) in *Collected Letters of John Maynard Keynes* Vol XIX, 299–301.

[17] In fairness to Keynes, he wrote this letter during the glory days of logical positivism when ontology was seen as the thoroughly disreputable branch of philosophy. It is a mark of his genius that he was swimming against this tide.

Newton's apple that yield data with which the actual values of the various quantities and factors in an equation or a formula can be filled, or is the idea of a non-existent object burdened with values, introspection, motives and expectations altogether absurd? If apples cannot be combined with motives, expectations and psychological uncertainties, but economics must nonetheless deal with them, must the apples be jettisoned, or is there a way of both having your apples and not having them by pretending that what you have is not an apple, but just like one? This, of course, is the route proposed by Adam Smith and the classical economists, including Marx, when they attribute an abstract character to labour in order to come up with an explanation of how different and incommensurable types of labour can nevertheless have the kind of homogeneity with which a theory of exchange can be stitched together. Smith writes:

> The greater part of people, too, understand better what is meant by a quantity of a particular commodity than by a quantity of labour. The one is a plain palpable object; the other an abstract notion, which, though it can be made sufficiently intelligible, is not altogether so natural and obvious.

In seeing the difference between the quantity of commodities and a (non-existent) quantity of labour as a matter of degree of naturalness or obviousness, Smith misses the fundamental ontological difference.

Mill's notion that there are apple-like objects that can be understood if separated from „disturbing causes" haunts Keynes reasoning. On the one hand, in this same letter he deems it „most important... to investigate statistically the order of magnitude of the Multiplier" as if the Multiplier were an apple-like object, but on the other, he sees that the „material" of economics, unlike that of the natural sciences, is „neither constant nor homogeneous". Is the Multiplier then an apple, or is it one of those materials that, due to being infected with motives, is neither constant nor homogeneous? Or is it neither? What should one do? Achieve homogeneity with Mill's device of eliminating as a „disturbing cause" anything obstructing homogeneity and constancy or admit that the inconstancy and heterogeneity of the „material" renders it unsuitable to scientific treatment. Keynes blinks.

Having posited as its singular objective the achievement of scientific status, much of modern economics has been about possible ways of overcoming this dilemma. The path was first set out in a systematic manner by Léon Walras, and, for the most part, remains the basic paradigm of economics. Walras offers his basic definition of „pure economics" in the Preface to the fourth and definitive edition of his *Elements of Pure Economics*[18] with deceptive clarity: „*Pure Economics* is, in essence, the theory of the determination of prices under a hypothetical regime of perfectly free competition." Note that, unlike Hendry, who is interested in understanding the *actual* transformation of substances, Walras wants to construct a theory about price determination under hypothetical circumstances. With this simple statement Walras makes it clear that the subject matter of economic theory, so to speak its apple, the determination of prices, is an imagined object that exists only under a

[18] Walras (1954)

non-existent hypothetical regime. And yet, he and his successors believe that one can do science without the apple.

In a passage of enormous significance for the development of modern economic theory, anticipating both Milton Friedman and Paul Samuelson, he sets out his claim with great clarity.

> … the physico-mathematical sciences, like the mathematical sciences in the narrow sense, do go beyond experience as soon as they have drawn their type concepts from it. From real-type concepts, these sciences abstract ideal-type concepts which they define, and on the basis of these definitions they construct *a priori* the whole framework of their theorems and proofs... the pure theory of economics ought to take over from experience certain type concepts, like those of exchange, supply, demand, market, capital, income, productive services and products. From these real-type concepts the pure science of economics should abstract and define ideal-type concepts in terms of which it carries on its reasoning. The return to reality should not take place until the science is completed and then only with a view to practical applications. Thus, in an ideal market we have ideal prices which stand in exact relation to an ideal demand and supply.

Remarkably, he adds a few lines later: „after that they (the scientists) go back to experience *not to confirm but to apply their conclusions.*" (Emphasis added.) He concludes his triumphant declaration with the dethroning of Ricardo and Mill and the crowning of mathematics:

> As to mathematical language, why should we persist in using everyday language to explain things in the most cumbrous and incorrect way, as Ricardo has often done and as John Stuart Mill does repeatedly in his *Principles of Political Economy*, when these same things can be stated far more succinctly, precisely and clearly in the language of mathematics?

The *a priori* framework of economic theory is, on this view, the product of two abstractions. The first abstraction produces „real-type" concepts from experience, to be followed by a second abstraction, which yields „ideal-type" concepts. Although Walras does not specify what each of these steps may involve, it seems reasonable to surmise from the text that the first of these steps involves some sort of isolation and/or reduction, something comparable to Mill's removal of disturbing causes, while the second one is ordinarily called idealisation. Both operations encounter serious difficulties when performed on objects of thought. In addition, Walras, as indeed all his successors leave us without an explanation as to how the representation of the ideal-type concept by numbers or symbols might be validated. When we substitute a number for that C, standing for capital in an equation, what exactly is that capital, and in what sense does that number comprise that C? The principal – often the only – known property of the ideal-type concept is that it bears a mathematically expressed relationship to something else. Unless we take as an axiom the view urged by Russell or Ryle that only quantifiable things exist, we must somehow account for intentionality. Are the numbers or symbols with which the ideal type is represented already embedded in experience, or do they materialise out of thin air in the course of the abstraction as the real-type and thereafter the ideal-type concepts emerge?

One could continue with an explication of the other great figures of the social sciences, such as Max Weber, John Stuart Mill or Carl Menger, but the common denominator is obvious: the methodology of the social sciences purports to create an ontologically objective object, but the object with which it ends up is not a suitable object for scientific treatment. At the same time, notwithstanding their unsuitability to science these surrogate objects, like allegories, metaphors and similes can reveal with great power some truth about a complex and confusing reality. There *is* relationship between supply and demand, between prices and quantities, and Alfred Marshall's two curves do reveal something important about that relationship. But price formation in the real world is far more complex and elusive, and, most importantly, far more a matter contingent on the interplay of an indeterminate number of factors than Marshall's curves allow. It is perfectly reasonable to build idealised models in the hopes of gaining some insight, but it is a great mistake to pretend that the subject matter of economics can be analogised to that of the natural sciences. Unlike the latter, in the famous words of Thomas Hobbes, it is „made with words".

Perhaps the most important consequence of this view of economics is that moral judgment is an inextricable part of its subject matter, that the construction of models and hypotheses without regard to that moral judgment leads not to the understanding of some sort of natural reality operating according to the laws of nature, but, rather, to an ideology parading as a science in order to be all the more convincing.

References

Anscombe, E.G.M. 1957. *Intention*. Oxford: Basil Blackwell.

Crane, T. 2013. *The objects of thought*. Oxford: Oxford University Press.

Geach, P.T. 1967. Intentional identity. *Journal of Philosophy* 64: 627–632.

Gilbert, M. 1989. *On social facts*. Princeton: Princeton University Press.

Hendry, R.F. 2008. Chemistry. In *The Routledge companion to philosophy of science*, ed. S. Psillos and M. Curd. London: Routledge.

Keynes, J.M. 2012. *The collected writings of John Maynard Keynes* Vol. XIX. Ed. E. Johnson and D. Moggridge. Cambridge: Cambridge University Press.

Mankiw, N.G. 2011. *Principles of economics*. Mason: Cengage Learning.

Meikle, S. 1955. *Aristotle's economic thought*. Oxford: Oxford University Press.

Pettit, Philip. 2008. *Made with words*. Princeton: Princeton University Press.

Priest, G. 2005. *Towards non-being*. Oxford: Clarendon Press.

Robbins, L. 1932. *An essay on the nature and significance of economic science*. London: Macmillan and Company.

Robinson, J. 1962. *Economic philosophy*. London: C.A. Watts.

Russell, B. 1919. *Introduction to mathematical philosophy*. London: George Allen and Unwin.

Searle, J.R. 1983. *Intentionality*. Cambridge: Cambridge University Press.

Smith, A. 1978. *Lectures on jurisprudence*. Oxford: Oxford University Press.

Walras, L. 1954. *Elements of pure economics*. Trans. W. Jaffé. London: George Allen and Unwin.

Wigner, E. 1960. The unreasonable efficiency of mathematics in the natural sciences. *Communications in Pure and Applied Mathematics* 13: 1–14.

Chapter 2
An Essay on Humble Economics

Łukasz Hardt

> *The man who claims for economic science much exactitude is a quack.*
>
> L. Robbins *(1927, 176).*

Abstract The purpose of this essay is to show that the departure from the Walrasian branch of neoclassical economics could lead to conceptualization of economics as a humble science, i.e. a science that is aware of its explanatory capabilities, but also convinced that discovering unambiguous regularities is not possible. Humble economics does not claim the power of formulating explanations which will not need further justification. It is based on an ontology of potentialities where one does not find laws, but capacities, tendencies, natures, and Aristotelian dynameis. It is based on the recognition that there is an irremovable mystery inherent in the functioning of the economy.

Faculty of Economic Sciences, University of Warsaw, and a member of the Monetary Policy Committee at the National Bank of Poland. This research was financed by a research grant within the framework of The National Programme for the Development of Humanities (NPRH) (grant no 2bH 15 0266 83).

Ł. Hardt (✉)
Faculty of Economic Sciences, University of Warsaw, Warsaw, Poland
e-mail: lhardt@wne.uw.edu.pl

P. Róna et al. (eds.), *Words, Objects and Events in Economics*, Virtues and Economics 6, https://doi.org/10.1007/978-3-030-52673-3_2

2.1 Introductory Remarks

Economics refers to laws, causes, regularities, mechanisms, or dependencies of a purely probabilistic nature. These issues are, however, the subject of hot disputes among economists themselves. Many of them believe that the explication of economic phenomena through their subsumption under laws does not make sense, while others would like laws in economics to resemble the status of almost universal laws of nature.[1] L. Walras was the first to claim that "[…] pure theory of economics is a science which resembles the physico-mathematical sciences in every respect" (Walras 1874/1984, 71), and then, almost 80 years after the publication of Walras' *Éléments d'Économie Politique Pure*, K. Arrow and R. Debreu offered a mathematically expressed proof of the existence of an equilibrium for a competitive economy. Their paper was published when the deductive-nomological model of explanation was at its peak of popularity. Therefore, it was only a step away from projecting economics as the physics of the social sciences. However, in the late 1960s economics started to become more and more diverse, and in the philosophy of science itself, the Hempel-Oppenheim model began to falter at its foundations. A discussion has begun on whether explaining empirical phenomena by subsuming them under general laws is appropriate at all.

Many important changes in economics took place in the 1970s. In this exceptional period a cognitive turn impacted economic science, key architects of the general equilibrium framework proved that macrophenomena are not easily reducible to micro ones, and thus the dominant mechanistic perspective has slowly begun to be questioned.[2] Economists such as H. Simon, D. Kahneman, A. Tversky, and V. Smith, in a sense, accomplished the unfulfilled dream of A. Marshall who wanted to make economics more rooted in the biological sciences.[3] K. Arrow himself said many years later:

> […] the very notion of what constitutes an economic theory will have to change. For a century, some economists have maintained that the biological is a more appropriate paradigm for economics than equilibrium models analogous to mechanics. […] economic theory may well take an analogous course (1995, 1617–1618).

The changes in neoclassical economics at the end of the twentieth century gained such a momentum that D. Colander in his insightful 2000 paper wrote that

[1] For the sake of simplicity, I equalize such laws to universal regularities (of the kind: if *X*, then *Y*) that are omnitemporally and omnispatially true, being at the same time characterized by a high level of necessity (Hardt 2017). Also, such laws are not equal to moral natural laws.

[2] Here I refer, among other factors, to the implications of the Sonnenschein–Mantel–Debreu theorem which states that aggregate demand functions that are built on individual preferences and other neoclassical postulates do not meet the weak axiom of revealed preferences (WARP). And since WARP is violated, then on the aggregate level it is impossible to prove the existence of an unique and stable equilibrium. Hahn (1995) saw the SMD theorem as the most devastating attack on micro-founded neoclassical economics.

[3] Here I refer to well-known words by Marshall: "The Mecca of the economist is economic biology rather than economic dynamics" (1898, 43).

neoclassical economics is dead.[4] The message of his text was that economics is so diverse and pluralistic that it makes no sense to use the term 'neoclassical economics' which is associated with the general equilibrium framework, and it is better to just talk about mainstream economics. The years that followed witnessed a further expansion of economic explanandum and explanans and, as a result, disputes erupted over whether the limits of economics are to be defined by its methods or a set of problems to which economics can be applied. Heated discussions about the status of neuroeconomics are a case in point.[5]

The above provokes the following question – was the Walrasian neoclassical economics only a short-term interlude in the development of economics which otherwise has a humble view of its research capabilities and communicates to the public that it can only acquire knowledge about contingent tendencies, causes, or natures of things, but not about universal and constant laws of nature? Classical economics was to a large extent such a humble science as it is illustrated by the well-known statement of N. Senior that "[The economist's] conclusions, whatever be their generality and their truth, do not authorize him in adding a single syllable of advice" (1836/1951, 2–3), or J.S. Mill's claim that economic laws are statements of tendencies only. Classical economists were much more careful in giving their opinions binding force than their neoclassical successors (Colander 2011).

The widespread use of formal models by neoclassical economists meant that the positivist understanding of science became less and less aligned with their research practice. Since modelling has become a *signum specificum* of economics, its methodological appraisal should include references to different ways of studying science, namely the ones subscribing to a model-based ideal of science, rather than to the well-established ideas of science built on laws of nature. This also makes it necessary to move away from Newton's deterministic mechanics and implies the need to face a multi-layered world in which unobservable factors play a key role.[6]

The purpose of this essay is to show that the departure from the Walrasian branch of neoclassical economics leads to the conceptualization of economics as a humble science, one that is aware of its explanatory capabilities, but also convinced that discovering unambiguous regularities is not possible. This humble economics does not claim the power of formulating explanations which will not need further justification. In other words, it is simply not possible to remove the mystery which is inherent in the functioning of the economy.

---

[4] It was even P. Samuelson, the so-called *Mr. Neoclassical Economics*, who claimed in 1976 that "To talk about neoclassical theories has certain implications. It is more appropriate to talk about mainstream economics" (in: Pizano 2009, 117). So, even he was somehow influenced by a changing spirit of economics in the 70's.

[5] See, e.g., a special issue of *Journal of Economic Methodology* (2010, no 2) dealing exclusively with the methodological status of neuroeconomics and its explanatory potential.

[6] We can even treat the unobservable as an anathema to many empiricist conceptions of science. Also, allowing the entry of unobservables into science makes metaphysics necessary in philosophical accounts of science (Chakravartty 2007, 16).

The essay is organized as follows. First, some preliminary insights concerning the way I understand a humble science are offered. Next, some reflections on the classical roots of modern economics are presented. It is shown that economists such as A. Smith, R. Ricardo, J.S. Mill, and A. Marshall were against dogmatism in economics. What follows is an analysis of metaphysics making its way back into philosophical reflection on economics. Next, I comment on D. Colander's essay on *Creating Humble Economists* (2016) and I propose that the lack of humility in economics cannot be helped by trying to persuade economists to behave like engineers (as it is suggested in Colander's paper) but rather by sticking with N. Cartwright's metaphysically rich statement that "Our most wide-ranging scientific knowledge [in this case knowledge about the economy] is not knowledge of laws but knowledge of the natures of things" (1999, 4). Conclusions follow.

2.2 Disentangling the Idea of a Humble Science

Science is about knowing the world. It is important, however, to realise that it is not an unlimited knowledge. Sticking to a position that science alone can explain every aspect of existence leads one inevitably to scientism which is a form of scientific dogmatism. Its proponents often believe that such a position can be made from within science. They are wrong. Statements about science are not scientific statements but rather meta-scientific ones and thus when reflecting on the scope of science one should benefit from insights from the philosophy of science.[7] And here, for instance, we can refer to K. Popper's claim that "It is important to realize that science does not make assertions about ultimate questions — about the riddles of existence, or about man's task in this world" (1978, 342). So, first of all, a humble science does not claim monopoly in providing us with knowledge on how the world works: "there is more to the physical world than has met the scientific eye" (Polkinghorne 1990, 88). Also, as L. Kołakowski wisely claims, "But although we cannot pierce the mystery and convert it into knowledge, our awareness that there is mystery is in itself important; although we cannot tear the veil from ultimate reality, we should know that such a veil exists" (2001, 10). Therefore, I disagree with people like P. Atkins (1994) who states that "There is nothing that cannot be understood" (1). But I shall not focus too much on studying the nature of general statements about science. My aim is rather to discuss the character of knowledge about the world that science (here economics) as such gives us.

[7] J. Lenox puts emphasis also on the fact that scientism contradicts itself: "The statement that only science can lead to truth is not itself deduced from science. It is not a scientific statement but rather a statement about science, that is, it is a meta-scientific statement. Therefore, if scientism's basic principle is true, the statement expressing scientism must be false. Scientism refutes itself. Hence it is incoherent" (2009, 43).

If our aim is to analyse the nature of scientific knowledge, then we should move towards epistemology.[8] Here questions of how we know things and how we can know that we know them emerge and they can be studied from various perspectives offered by numerous theories of knowledge. The traditional dilemma of knowing also applies: are we able to know the world for certain (absolutism) or can we just know it from a mere perspective (relativism) and, therefore, nothing is known with absoluteness or certainty (scepticism)? Simply speaking there is a long tradition in Western thought to see this very dilemma as a matter of all or nothing, as a choice between dogmatism and scepticism. However, there is also a short history of thought that searches for some middle ground in this debate. My take on a humble science subscribes to this latter tradition.

Since the subject of this paper is economics, there is a particular justification to refer to Hume while studying the issue of epistemological humility: apart from being one of the most important philosophers studying the nature of human understanding, he was also a crucial source of inspiration for Adam Smith. At the very end of *An Enquiry Concerning Human Understanding* we can find the following description of human beings as inherently epistemologically humble: "[People] must act and reason and believe; though they are not able, be their most diligent enquiry, to satisfy themselves concerning the foundation of these operations" (Hume 1748/2004, 104). In Hume's case his epistemological humility is strongly intertwined with his ontological humility: "Nothing in the world is perpetual. Everything, however, seemingly firm, is in continual flux and change. The world itself gives symptoms of frailty and dissolution" (1777/2006, 603). As Holland (2013, 31) writes, "[…] no one in the European philosophical tradition has a keener sense of the limits of human knowledge than David Hume". Therefore, knowing something for certain is simply impossible for him and the only thing one can achieve is to have "a degree of belief, which is sufficient for our purpose" (1740/2000, 122). However, a belief which would be free from any doubts is hardly possible: "Belief, being a lively conception, can never be entire, where it is not founded on something natural and easy" (ibid.). Interestingly, Hume claims that beliefs are somewhere in-between knowing something for certain and knowing something with a given degree of probability only. However, Hume's beliefs are subjective, and they are based on impressions about the causal structure of the world. Therefore, they are always subject to error, since we only *feel* a certain necessity in the causal connection between empirical states.

Hume's introduction of beliefs into epistemology was followed by various thinkers offering this category a legitimate place in modern philosophy of science. Importantly, for instance, Kant in his *Critique of Pure Reason* offers three categories of human understanding, namely opinion, knowledge, and belief, and thus writes:

[8] It does not, however, mean that we are to commit a kind of 'epistemic fallacy', namely a reduction of ontology to epistemology.

> Holding for true, or the subjective validity of a judgment in relation to conviction (which is, at the same time, objectively valid), has the three following degrees: opinion, belief, and knowledge. Opinion is a consciously insufficient judgment, subjectively as well as objectively. Belief is subjectively sufficient, but is recognized as being objectively insufficient. Knowledge is both subjectively and objectively sufficient (1781/1996, 749).

Here, as in Hume, belief is subjective. Nevertheless, Kant is a little less pessimistic than Hume since what we know about a given thing is a result of an interaction between our experience of this very object and its "thing-in-itself". Instead of delving further into Kant, I shall move to Charles Peirce who (together with W. James) founded pragmatism - a philosophical movement claiming that "the rational justification of scientific beliefs ultimately depends on whether the method generating the beliefs is the best available for advancing our cognitive goals of explanation and precise prediction" (Almeder 2014, 103). Here is Peirce's famous passage about the role of beliefs:

> Doubt is an uneasy and dissatisfied state from which we struggle to free ourselves and pass into the state of belief; while the latter is a calm and satisfactory state which we do not wish to avoid, or to change to a belief in anything else. [...] the irritation of doubt causes a struggle to attain a state of belief. I shall term this struggle *inquiry* (1877, 66–67).

Such world view is often criticized for rejecting the very notion of truth and focusing solely on the utility of beliefs (see, e.g., Rorty 1995). I think that we may let pragmatists defend so strongly the utility of beliefs but at the same time claim that at least some of these beliefs are true. Also, true beliefs do not have the status of Humean regularities, e.g., believing that lower interest rates should lead to higher investments can still be true even if in a given situation (say in Poland in 2019) this is not the case.[9] What is very important in pragmatists' stance on the character of knowledge is that the way beliefs are formed matters for their justification.

I hope that it is now clear that insights given by Hume, Kant, and Peirce can be seen as supporting the ideal of a humble science, one that accepts fallibility of some of its claims. Or, to be more precise, a humble science does not reject the possibility that some of its statements may just be wrong. The character of humble science can also be illustrated by contrasting it with its opposite, namely a science characterized by vices, including arrogance, vanity, conceit, egotism, grandiosity, pretentiousness, snobbishness, impertinence (presumption), haughtiness, self-righteousness, domination, selfish ambition, and self-complacency (Roberts and Wood 2003, 258). If the way we practice science matters for its success (as pragmatists claim), one may ask whether humble scientists are always to opt for a humble science. Here we are approaching virtue epistemology where many authors claim that intellectual humility serves as a prerequisite for a humble science. Hazlett (2012, 220) states that intellectual humility is "the disposition not to adopt epistemically improper higher order epistemic attitudes". Also, but in a slightly different context, Whitcomb

[9] This can be so because inference can be of a purely probabilistic nature, disturbing causes can be at place, etc.

et al. (2015) claim that being a humble scientist means to know one's self-limitations and thus being open to criticism. Roberts and Wood (2003, 272) nicely describe a humble researcher by portraying her antithesis:

> The intellectually vain person is overly concerned with how he 'looks' to the people who count: he wants to impress, and is very concerned not to look silly at conferences and in front of his bright students. This concern may incline him not to admit, and maybe not even to notice, when someone has raised a good objection to his views. It may also incline him to fudge arguments when he thinks he can make them look good enough to get away with. The intellectually vain person may be genuinely concerned to accomplish intrinsic epistemic ends: to figure out what's what and to give his students a good education. But he also has the extrinsic concern to look good intellectually, and we are saying that this is in general an epistemic liability. By contrast, the lack of concern to look good frees the intellectually humble person to pursue intellectual goods simply and undistractedly (think of G. E. Moore).

Saying that „everyone nowadays is, I take it, a fallibilist about scientific theories" (Worrall 1989, 268) does not mean that we cannot approach the truth about the world and that there are no places where Humean regularities can always be true. They can hold, for instance, in theoretical models. What I claim is that vices like pride, haughtiness, and conceit can block any scientific progress since one can just conclude that he knows everything about the world and no further scientific inquiry is necessary. In other words, pride leads to intellectual laziness. But humility, on the contrary, makes people curious about the way the world works and therefore "it is humility that makes men as angels" (St. Augustine), and hence as McCloskey (2006) stresses when citing T. Merton: "Humility is a virtue, not a neurosis. A humility that freezes our being and frustrates all healthy activity is not humility at all, but a disguised form of pride" (1956, 55). One can also refer to Saint John Paul II's lecture at the Jagiellonian University in Cracow in 1997: "Reason should act and be active in the spiritual climate of moral virtues, like honesty, courage, humbleness, and genuine concern for human beings" (St. John Paul II 1997/2006, 988). And in *Fides et Ratio* he adds: "Human wisdom refuses to see in its own weakness the possibility of its strength; yet Saint Paul is quick to affirm: 'When I am weak, then I am strong' (2 Cor 12:10)" (no 23).

At the very beginning of this section I wrote that a humble science is a kind of a middle ground between dogmatism and scepticism. Now, I can add to the above that intellectual humility can be treated as an Aristotelian mean state between absolutism of certitude and timidity of incertitude. Therefore, humility is not only a moral virtue but also an epistemic one. In any case, they are intertwined since being a humble person makes it easier to be a humble scientist who can practice a humble science. In the following section I will look at classical economists and show that economics they practiced can be treated as an example of a humble science.

2.3 Classical Economics as a Humble Science

When focusing on classical economics, one usually starts from the analysis of Adam Smith's works. However, in order to understand Smith, we should take note of D. Hume and his reflection on the character of knowledge economics offers us.[10] We did it partially in the previous section, but now we would like to be more precise and focus solely on his economics.[11] Hume used a lot of thought experiments in his economic research. As Massey (1991) comments "[…] perhaps no other philosopher has conducted his thought experiments with the degree of care and sophistication that Hume bestowed on his" (293). Also, Schabas concludes her paper on models in Hume's thought by writing that "I have not yet found such a concentration of examples [of thought experiments] since the ones devised by Hume" (2008, 168). What then do Hume's thought experiments look like? Here I would like to refer to one of his most important experiments, namely the analysis he presented in the *Essay of Interest* (1758/1993, 181):

> For suppose, that, by miracle, every man in Britain should have five pounds slip into his pocket in one night; this would much more than double the whole money that is at present in the kingdom; and yet there would not next day, nor for some time, be any more lenders, not any variation on the interest (emphasis added).

Referring to the "miracle" means that we are dealing here with an imaginary situation, a kind of thought experiment, and not a precisely defined and isolated model.[12] Moreover, this magical appearance of additional money in the pockets of British subjects indicates that Hume analyses solely the influence of monetary variables on the economy – only they have meaning, and other factors are either fixed or absent (ceteris paribus). This is important because it served him to show the neutrality of money. Explaining empirical facts through thought experiments presupposes that the mechanisms described in them have their counterparts in the economy. It is difficult to find in Hume's thought anything that better supports this thesis than the following words:

[10] There is a huge amount of literature on Hume-Smith connection and presenting it in details is not necessary here, however, I would like just to cite Smith from his letter written to W. Strahan shortly after Hume's death: "I have always considered him, both in his lifetime and since his death, as approaching as nearly to the idea of a perfectly wise and virtuous man, as perhaps the nature human frailty will permit" (Smith 1776, cited in: Rasmussen 2018, 51).

[11] Writing about Hume's economics does not mean that he can be undoubtedly called an economist. He was definitely a moral philosopher and only later started to be described as an economist (Schumpeter 1954/2006, 120).

[12] Although models and thought experiments are similar in many respects, one should notice also important differences. First, models are more decoupled from their targets than thought experiments. Second, such experiments (at least in Hume's interpretation) should refer to empirical targets that are at least potentially possible; models, on the other hand, can describe fairy-tale worlds. Third, thought experiments' insights are often illustrated by some references to empirical domains, but the ones of models are usually discussed in purely theoretical schemes or pictures (Schabas 2008).

> Now it is evident that the same causes which would correct these exorbitant inequalities, were they to happen miraculously, must prevent their happening in the common course of nature, and must forever in all neighboring nations, preserve money nearly proportional to the art and industry of each nation (Hume 1758/1993, 191).

The first fragment of the above paragraph from Hume's work refers to the situation of the thought experiment, whereas the second part points out to empirical reality, hence the use of the phrase "nearly proportional", not simply "proportional". The above shows that Hume willingly went from analysis in terms of a thought experiment to the study of the surrounding empirical phenomena. In his text *On Public Credit* (1758/1993) we can find an interesting reference to the then Great Britain: "[the situation] to which Great Britain is visibly tending" (211).[13] What is always true in the thought experiment, because it happens "miraculously", namely in isolation from other potentially disturbing causes, in reality takes place in a tendency only. To be more precise, a given cause produces a tendency to a particular result and not necessarily the very result itself. Certainty can only be in place in thought constructs, strictly axiomatized and defined. It is in these model worlds that there are universally true relationships and laws. In his *Treatise,* while accounting for the nature of human knowledge, he writes:

> Here remain, therefore, algebra and arithmetic as the only sciences, in which we can carry on a chain of reasoning to any degree of intricacy, and yet preserve a perfect exactness and certainty. It is impossible for the eye to determine the angles of a chiliagon to be equal to 1996 right angles, or make any conjecture, that approaches this proportion (Hume 1740/2000, 51).

Before commenting on this passage, I would like to quote below the following statement from Mill's 1836 essay *On the Definition of Political Economy*:

> The conclusions of geometry are not strictly true of such lines, angles, and figures, as human hands can construct. But no one, therefore, contends that the conclusions of geometry are of no utility, or that it would be better to shut up Euclid's *Elements*, and content ourselves with 'practice' and 'experience' (Mill 1836/2008, 46).

Both Hume and Mill state that in isolation, i.e., in a given model (Mill) or in a particular thought experiment (Hume), certainty is possible and that true statements in these theoretical worlds become approximations of truth if they are related to empirical facts – a triangle drawn by the author of this essay is to have only approximately 180°, and this is even more striking in a hypothetical case of a chiliagon. So, we will never have universally binding and precise laws that are also context independent.

Now, and before moving to Smith, it is quite noteworthy that some interesting reflections on humility and pride can be found in Hume's *Treatise*. For instance, in the section titled *Of Greatness of Mind* Hume starts his analysis by clearly stating that "An excessive pride or overweening conceit of ourselves is always esteemed vicious, and is universally hated; as modesty, or a just sense of our weakness, is esteemed virtuous, and procures the good-will of every-one" (Hume 1740/2000,

[13] This sentence did not appear in the first edition of Hume's essay.

378). Earlier in the text he claims that "Humility exalts; but pride mortifies us" (ibid., 193). Although these words are put forward in the sections of his work which deal with morals and not human understanding as such, one can apply them to the issue of the nature of human knowledge in view of Hume's analysis presented in his *Enquiry Concerning Human Understanding*. In Hume's *Enquiry* we can find a clear statement denying the possibility of reaching ultimate knowledge, while at the same claiming that it is modesty which eases researchers to be humble in assessing the character of knowledge they possess about the world: "Hence we may discover the reason why no philosopher, who is rational and modest, has ever pretended to assign the ultimate cause of any natural operation, or to show distinctly the action of that power, which produces any single effect in the universe" (Hume 1748/2004, 17). Here reference is made to philosophers but nowadays we can also apply it to economists. Hume's vision of human knowledge then is very close to the one we described earlier as a humble science.

As far as Adam Smith's ideas concerning the nature of laws in economics are concerned, one may say that they are somehow contradictory. However, they are such only on the surface and do not give ground to another *Das Adam Smith Problem.* Since Smith was hugely influenced by Newton, he shared with him a mechanistic world view in which one has structures, mechanisms, and never changing relationships. In other words, both conceived of a reality in which omnispatially and omnitemporally constant conjunctions are possible. Take, for instance, the following insight from The *Theory of Moral Sentiments*: "Human society, when we contemplate it in a certain abstract and philosophical light, appears like a great, an immense machine, whose regular and harmonious movements produce a thousand agreeable effects. (1759/2013, 280). In Smith's *Essays on Philosophical Subjects* there is another, clearly stated observation: "The universe was regarded as a complete machine, as a coherent system, governed by general laws, and directed to general ends" (1795/1980, 113). On the other hand, we can also find Smith's statements that seem to be totally antinomous to the above ones. In *The Theory of Moral Sentiments,* he writes: "In the great chess-board of human society, every single piece has a principle of motion of its own" (1759/2013, 204). It seems therefore that, we do not have universal rules of human behaviour but each human subject has its own principles of conduct. How then can we reconcile the two seemingly contradictory views on the nature of knowledge about the economy? Here Smith himself can help us with the following statement:

> Systems in many respects resemble machines. A machine is a little system, created to perform, as well as to connect together, in reality, those different movements and effects which the artist has occasion for. A system is an imaginary machine invented to connect together in the fancy those different movements and effects which are already in reality performed. The machines that are first invented to perform any particular movement are always the most complex, and succeeding artists generally discover that, with fewer wheels, with fewer principles of motion, than had originally been employed, the same effect may be more easily produced. The first systems, in the same manner, are always the most complex, and a particular connecting chain, or principle, is generally thought necessary to unite every two seemingly disjoined appearances: but it often happens, that one great connecting principle is afterwards found to be sufficient to bind together all the discordant phenomena that occur in a whole species of things (Smith 1795/1980, 66).

Is not Smith just writing about modelling economic phenomena? His "imaginary machines" are models that are used to "discover principles", including those governing economic systems. In his interpretation of the above cited excerpt Condorcet concluded that "in all the arts, the truths of the theory are necessarily modified in practice" (Condorcet 1804, 292).[14] So, in models or thought experiments one can have perfect and unchangeable regularities, but once we refer them to empirical domains we only have some beliefs or "some degrees of certainty to which we may hope to attain" (ibid., 138). Referring to a more recent example, one can say that in the model of perfect competition price is always equal to the marginal cost of production, but outside the model, when we appeal to the real economy, p = MC does not, in fact, obtain. Economic models do not give us certain knowledge about the world. Thus, the most we can do is to humbly make inferences from models to their targets.

J. S. Mill was another nineteenth-century economist who questioned the possibility of formulating precise and universally binding laws of economics. Instead, he proposed the following definition of economic laws: "All laws of causation, in consequence of their liability to be counteracted, require to be stated in words affirmative of tendencies only, and not of actual results" (Mill 1843, 523). Mill also wrote about the idea of "abstract truth", namely the truth of theoretical claims insofar as they are placed in abstract places (models). However, "the conclusions correctly deduced from these assumptions [models], would be as true in the abstract as those of mathematics; and would be as near an approximation as abstract truth can ever be, to truth in the concrete" (Mill 1836/2008, 49). Interestingly, Mill's tendencies are acting powers that make things happen (ibid., 56). I shall elaborate on this point in the following section.

Now, before closing our study on classical economists, I shall recall A. Marshall who is rightly treated as one of the founding fathers of neoclassical economics but whose economics did not lose its classical roots (Colander 2011). One can even argue that the Marshallian branch of neoclassical economics is closer to the ideas of Smith, Ricardo, and Mill than to the Walrasian branch of neoclassicism. This is particularly true of Marshall's insights concerning the nature of economic laws. Suffice is to quote some select views of his here: "If the subject matter of a science passes through different phases of development, the laws of the science must have a development corresponding to the things of which they treat" (Marshall 1890, 65); "Every cause has a tendency to produce some definite result if nothing occurs to hinder it. Thus gravitation tends to make things fall to the ground: but when a balloon is full of gas lighter than air, the pressure of the air will make it rise in spite of the tendency of gravitation to make it fall" (Marshall 1920/2013, 26); "there are no economic tendencies which act as steadily and can be measured as exactly as gravitation can: and consequently there are no laws of economics which can be compared

[14] M. de Condorcet was one of the most important interpreters of Smith's ideas (Rothschild 2001). Although he did a lot of research in theory of probability, he was conscious that economic systems are so complex that interplays between various events are not of a probabilistic nature, and thus he often referred to beliefs as being somehow in-between knowledge and probability.

for precision with the law of gravitation" (ibid.). So, simplifying a bit, not only does each economy have its own set of laws but these laws are statements about tendencies only. Therefore, not being humble in assessing our abilities of understanding economic systems would be just wrong.

In what follows, I will come back to Mill's tendencies but in a metaphysically richer way. In doing so I will claim that while assessing the nature of our knowledge about the economy we cannot abstract from the very character of the economic realm. In other words, ontological fundamentals of economic systems matter for the scope and certitude of our knowledge about them. Or, as Lawson 2015 claims "[…] method must fit with the nature of its object" (13), and thus, for instance, Róna states that "clarity on the ontology of the subject matter of the social sciences seems urgently needed" (2018a, 5).

2.4 Metaphysics of Economic Systems

If we would like to describe an economic reality, we should ask how its organizing parts come to be. More fundamentally, a study into the nature of specific elements of this reality is necessary. It is claimed here that "all features of reality can be viewed under the aspect of their being" (Lawson 2014, 19). And scientific ontology can be understood as primarily interested in investigating the natures of particular economic objects. It differs from philosophical ontology which deals with general aspects of being.[15] As I show in my recent book on *Economics without Laws* (2017), and what I have just indicated with respect to classical economics, the economic world is not governed by universal laws such as *whenever A, then (always) B*, but rather by such entities as powers, mechanisms, tendencies, and structures. Such a world is a world of potentialities only. Also, speaking metaphorically, such a world is a world where its constituting parts interact "chemically" and not "mechanistically".[16] In such a realm, macro phenomena emerge from their micro

[15] M. Bunge, for instance, describes this difference as follows: "Ontology can be classed into general and special (or regional). General ontology studies all existents, whereas each special ontology studies one genus of thing or process physical, chemical, biological, social, etc. Thus, whereas general ontology studies the concepts of space, time, and event, the ontology of the social investigates such general sociological concepts as those of social system, social structure, and social change" (Bunge 1999, 200). Our scientific ontology, and economic ontology in particular, is an example of Bunge's special ontology.

[16] Here I refer to Mill's analysis presented in his chapter *On the Composition of Causes* in *A System of Logic*. According to Mill, when causes interact chemically then we cannot use the method of isolation and thus the very ontological foundations of classical (and neoclassical) economics need to be questioned. In Mäki's words, "This is the question of whether the causes of economic phenomena combined 'mechanically' or 'chemically', to use Mill's phrases. When causes combine 'mechanically', their effects can be 'added up' like vectors […]. On the other hand, when causes are combined 'chemically', some qualitative novel, emergent outcomes ensue" (1992, 349).

foundations rather than being reducible to their constituting parts. So, as Lawson puts it:

> [...] these or related notions [law-like statements in economics] must be conceived in terms of potentials; as potentials that may or may not be expressed, and if expressed that may or may not be actualized because of countervailing tendencies [...] (Lawson 1997, 106).

And before he also writes that:

> [...] science aims at uncovering causal factors, that is, it is concerned with identifying structures, mechanisms and the tendencies they ground, which produce, govern or facilitate phenomena at a different level. And if the aim of science is to illuminate structures that govern surface phenomena then *laws* or *law-statements* are neither empirical statements (statements about experiences) nor statements about events or their regularities (whether unqualified or subject to *ceteris paribus* restrictions), but precisely statements elucidating structures and their characteristic modes of activity (ibid., 24; italics in original).

Three ideas from the above statements by Lawson are important for our purpose, namely those that treat about tendencies, characteristic modes of activities, and ceteris paribus clause. So, let me begin with tendencies. They can be defined by referring to N. Cartwright's philosophy of potentialities, i.e.: "Substituting the word 'capacity' for Mill's word 'tendency', his claim is exactly what I aim to establish in this book [...]. I suggest that the reader take my 'capacity' and Mill's 'tendency' to be synonymous" (Cartwright 1989, 170). The following may serve as an example: Mill would claim that lower interest rates produce a tendency of investments to rise. Cartwright would rather say that lower interest rates carry the capacity to cause higher investments. Or, to use Cartwright's favourite example, saying that aspirin relieves headaches should be read as claiming that aspirin has the capacity to relieve headaches.[17] Now, we come closer to the second idea underlined above by Lawson, namely things' characteristic modes of activity. In Cartwright's terms one should just say that the very nature of *A* is to produce *B*. For instance, the nature of lower interest rates is to make investments higher. But, as in the case of aspirin, a particular capacity may be dormant, or its functioning may be offset by some disturbing factors. So, one may have low interest rates without a corresponding shift in investment. No universal laws please, but only natures are real, and they make economic events happen: "Aristotle's notion of nature is far more suitable than the concept of *law*, *regularity* and *occurrent property* to describe the kind of knowledge we have" (Cartwright 1999, 78; original italics). Claiming the contrary would lead one to fundamentalism which we should oppose.

What now about ceteris paribus clause? Would it not suffice as a basis for a humble economics? It would not as long as it is understood merely as 'other things being constant' restriction. Also, such laws are either empirically false (because disturbing factors occur) or they are trivially true (if they are understood as purely analytical statements). Countless papers have been written on how such laws can be

---

[17]As the author of *Metaphysics* put it: "Something is said to be a capacity [potentiality, power] when it is a starting-point of movement or change either in another thing or in a thing insofar as it other" (Aristotle 2016, 83).

defended. Here I would refer to just two such defences that interpret ceteris paribus laws as humble statements about the world. First, just before studying ceteris paribus clause in the third chapter of his *Principles* A. Marshall wrote the following: "And following our definition of an economic law, we may say that the course of action which may be expected under certain conditions from the members of an industrial group is the normal action of the members of that group relatively to those conditions" (Marshall 1920/2013, 28). Marshall talks here about a *ceteris normalibus* clause which says that a particular statement is only true when conditions are normal. For instance, a rise in unemployment *ceteris normalibus* lowers inflation. Such a statement is only true in a particular New Keynesian model where one can have a non-flat Philips curve. Outside such a model we can only believe that such a relation can hold imperfectly. And here we are somehow coming back to Cartwright's understanding of economic realm mentioned above where *ceteris normalibus* laws simply become statements of the following kind: *ceteris normalibus A* leads to *B*, and it means that the nature of *A* is to produce *B*. So, and again, we have a humble way of interpreting laws of economics. As Cartwright nicely describes it:

> There is a tendency to think that all facts must belong to one grand scheme, and, moreover, that this is a scheme in which the facts in the first category have a special and privileged status. They are exemplary of the way nature is supposed to work. The others must be made to conform to them. This is the kind of fundamentalist doctrine that I think we must resist (1994, 316).

We live in a dappled world where we do not have all encompassing theories that give us a sense of certitude. Here I agree with Rodrik (2015, 17), who humbly says, "The correct answer to almost any question in economics is: it depends". But even if we have an appropriate set of theoretical statements crafted for a given empirical domain, we will then be able to describe merely capacities, natures, and powers of the objects under investigation. The above reflection is of key importance for my thesis about the necessity for humility in building economic explanations. Since models produce beliefs that economists have about the world and if they are credible when the structure of a given model is close to its empirical target, it turns out that, for instance, economic laws (even if understood as statements about tendencies only) which give an accurate description of highly developed countries do not have to be adequate in explaining phenomena in less developed countries. Thus, it is not only the case that context is everything but even in a given context we are unable to guarantee that if *A, then (always) B* rules, and hence capacities are (nearly) everything, too.

2.5 D. Colander's Plea for Creating Humble Economists

In an important paper on the current state of economic profession D. Colander (2016) claims that the crucial problem economics is facing is "lack of humility" (737). Therefore, he writes that "we economists have a tendency to convey more scientific certainty in our policy positions than the theory and evidence objectively would allow" (ibid.). Later he proposes the way for having more humble economists. His solution is to treat economists not as applied scientists but rather as engineers. Colander's argument for such a solution rests on the assumption that the goal of science "is finding the truth" (738) while that of engineering is "by nature applied" (739). And he cites Koen's statement that the engineering method is "The strategy for causing the best change in a poorly understood or uncertain situation within the available resources" (ibid.). What matters is also the fact that Colander's understanding of truth is rather straightforward: "a scientific truth is timeless" (ibid.). Next, he claims that most economists are actually engineers and that they are by nature more humble than pure scientists.

Colander's arguments are roughly similar to the ones of B. Bernanke, who, while heading the FED, in his Princeton 2010 speech claims that in fact we have three areas we should focus on, namely economic science, economic engineering, and economic management. In his opinion we have more problems with the two latter concepts than with the former one. I do not entirely agree with either Colander or Bernanke, but I find the former FED's chair arguments a little more appealing since I also think that we do not have problems with economics as such but rather we have some problems with the correct understanding of the claims it makes. However, the very act of correct understanding of economics' insights does not belong to economic engineering either in Bernanke's or Colander's sense. In other words, Bernanke is wrong in claiming that economic science offers theoretical and empirical generalizations, and Colander is similarly wrong in stating that economics should just move more towards engineering and not science. What I would like to stress, and what should hopefully be clear from the insights suggested in previous sections, is that there can be a science, also economic science, which does not offer us eternal and unchangeable strict generalizations but rather informs us about capacities of potentially causative factors. It can inform us about some strictly defined behaviours only insofar as they appear in closed theoretical models. Such rules of behaviour usually do not perfectly hold once we move from models towards their targets, instead of generalizations we rather have some beliefs about real economies. Moreover, these beliefs can be treated more like *ceteris normalibus* laws or, to be even more precise, as mixed cn-cp laws, e.g., *ceteris normalibus* (for ideal markets, i.e., in some model conditions) and ceteris paribus (having other variables unchanged), a decrease in interest rate leads to an increase in investments. Also, a particular *ceteris normalibus* clause can be understood not only as being synonymous to a given model condition but rather as a metaphysically rich conditioner, namely *the nature of A is to produce B*. Here we are close to new Aristotelianism. It is, therefore, not only the case that science without laws is possible. Science which does not offer timeless truths is possible as well. And so, paraphrasing Giere (2000, 523), what has traditionally been interpreted as laws of nature describing various economies, turn out to be merely statements describing the behaviour of theoretical economic models. One who realizes this, is very close to a humble economics.

Colander's insistence on the fact that the most important problem economics is facing is lack of humility in economists' behaviour is definitely true. Also, my claim that we need a humble economics, namely economics conscious of the kind of knowledge it produces (as it was discussed earlier) necessarily implies that we need economists who understand economics in such a way. My disagreement with Colander's proposal is that we do not have to transform economists into engineers to achieve such a change. We should not abandon science in general, and economics in particular, and the way forward is towards a humble science, and in our case a humble economics.

2.6 Conclusions

There is an ongoing debate concerning the state of modern economics. Some accuse economics of "mistaking beauty, clad in impressive-looking mathematics, for truth" (Krugman 2009); others point to the fact that we do not have any problems with the science of economics but rather with economic engineering and management (e.g., Bernanke 2010); some emphasise the fact that we have problems not with economics as such but with economists and their lack of humility (Colander 2016). In the reflections presented in this paper I also focused on the lack of humility but in a somewhat different sense. The focus here was on the problem that flies in the face of a great number of economists in many areas of economics, namely their dream of finding universal laws governing economic systems. However, these laws do not exist. As Róna (2018b, 189) claims, for instance: "The proper subject of economics, therefore, cannot be the search for nature based law-like generalities with reliable predictive and explanatory power, because there are no such generalities". Although many economists would agree that laws they discover do not have the status of laws of nature, they still have many problems in rejecting the worldview which sees the world as being governed by such laws. However, and again, laws of nature do not rule the world and thus searching for scientific laws that do is futile. What we can aim for is to identify capacities, causes, mechanisms, tendencies, and some imperfect empirical regularities. Unchangeable and omni-spatio-temporally binding laws can only apply inside theoretical models. Claiming that they are to hold in the same way beyond models is mistaken. In other words, taking the model mistakenly for its empirical target may easily lead one to a form of scientific dogmatism. As Rodrik nicely puts it: "there is no such thing as *the* model, but always it is *a* model" (2015, 43). So, explaining the workings of economic systems by using models is inherently imperfect. Said differently, inferences from models to their targets should be made with due humility.

Although we still face the problem of economics' failure to adequately understand the status of its claims and of assigning too much certitude to them (as discussed above), there are many signs which indicate that the economic profession is now turning (though slowly) in the right direction. One example is what is often called an empirical turn in economics which means that more and more papers are applied and focused on data rather than on theory only. Therefore, one can claim that there is a diminishing risk of taking models' claims outside their proper domains. But still, such a turn in economics does not make economic science less prone to criticism but rather changes what is criticized and how. Studies based on randomized control trials can serve as an example. And there we have a problem of external validity. If you want RCTs results to apply elsewhere in any simple way, you commit the very same error as authors of theoretical models in economics: they want their insights to work beyond their proper domains. Since experiments (including RCTs) are models, their results must be interpreted with due humility. Writing about RCTs Deaton and Cartwright (2017) claim the following: "We often learn much from coming to understand why replication failed and can use that knowledge, in looking for how the factors that caused the original result might operate differently in different settings". In other words, causes of various economic phenomena identified in empirical work should be given careful (humble) interpretation, e.g., they are only INUS conditions for particular results.[18] Let me explain. Imagine that a lower interest rate causes investment to rise. What should be accounted for with regard to this statement is the presence of firms that would like to invest but need cheap capital, and some regulatory rules that allow for firms' investments. Together these factors are unnecessary but sufficient to the increase of investment rate (since many other events certainly could lead to higher investment, e.g., a given firm may receive a subsidy from the government, etc.). Within this very set of three events, the lowering of interest rate is an insufficient (since lowering of the interest rate only, without firms willing to invest, for instance, would not cause higher investment) but non-redundant (because higher investment would not occur without it, ceteris paribus) part of a condition which is itself unnecessary but sufficient for the occurrence of the result. Therefore, a lower interest rate is an INUS condition for higher investment. So, we are again facing the issue of proper interpretation of knowledge about the world the empirical experiment gives us. Here again a correct way to tackle it is to treat it humbly. With regard to the above example the most one can claim is that the nature of lower interest rate is to stimulate investment. This is the way Cartwright in her writings seems to interpret INUS causality, including the one identified via RCTs.

As we have just seen, although empirical turn is a step in the right direction for economics, it does not alone guarantee a move towards a humbler economics. One may say that it can be treated only as an INUS condition for making economics humbler. Other developments such as making economics more pluralistic and hence context sensitive may also (potentially) make it humbler. Yet, even in thus modified economics some researchers may still give too much certitude to the insights they produce. Therefore, for a genuine movement towards a humble economics one needs economics based on an ontology of potentialities where one finds not laws, but capacities, tendencies, natures, and Aristotelian dynameis. In the very least I hope that philosophy of economics will contribute a bit to the occurrence of such a change by making economics more self-critical.

---

[18] INUS causality is due to Mackie (1974), and INUS clause stands for Insufficient but Non-redundant parts of a condition that is itself Unnecessary but Sufficient for a contribution to the outcome.

References

Almeder, R. 2014. Pragmatism and science. In *The Routledge companion to philosophy of science*, ed. S. Psillos and M. Curd, 103–111. Oxon: Routledge.

Aristotle. 2016. *Metaphysics*. Cambridge: Hackett Publishing Company.

Arrow, K.J. 1995. Opinion concerning the future of science. *Science*. Issue published on 17/03/1995.

Atkins, P. 1994. *The creation revisited*. Harmondsworth: Penguin.

Bernanke, B. 2010. Implications of the financial crisis for economics. Speech delivered at the Princeton University. *BIS Review* 123.

Bunge, M. 1999. *Dictionary of philosophy*. Amherst: Prometheus Books.

Cartwright, N. 1989. *Nature's capacities and their measurement*. Oxford: Clarendon Press.

———. 1994. Fundamentalism vs. the patchwork of laws. *Proceedings of the Aristotelean Society* XCIV: 279–292.

———. 1999. *The dappled world: A study of the boundaries of science*. Cambridge: Cambridge University Press.

Chakravartty, A. 2007. *A metaphysics for scientific realism: Knowing the unobservable*. Cambridge: Cambridge University Press.

Colander, D. 2000. The death of neoclassical economics. *Journal of the History of Economic Thought* 22 (2): 127–143.

———. 2011. How economists got it wrong: A nuanced account. *Critical Review: A Journal of Politics and Society* 23 (1–2): 1–27.

———. 2016. Creating humble economists. In *The Oxford handbook of professional economic ethics*, ed. G. DeMartino and D.N. McCloskey, 737–749. Oxford: Oxford University Press.

de Condorcet, M. 1804. *Esquisse d'un Tableau Historique*. Paris: Henrichs.

Deaton A., and N. Cartwright. 2017. Understanding and misunderstanding randomized controlled trials. NBER working paper, no 22595.

Giere, R.N. 2000. Theories. In *A companion to the philosophy of science*, ed. W.H. Newton-Smith, 515–524. Oxford: Blackwell Publishers.

Hahn, F. 1995. *A critical essay on modern macroeconomic theory*. Cambridge: MIT Press.

Hardt, Ł. 2017. *Economics without laws. Towards a new philosophy of economics*. Cham: Palgrave Macmillan.

Hazlett, A. 2012. Higher-order epistemic attitudes and intellectual humility. *Episteme* 9 (3): 205–223.

Holland, N.J. 2013. *Ontological humility: Lord Voldemort and the philosophers*. Albany: SUNY Press.

Hume, D. 1740/2000. *A treatise of human nature*. Oxford: Oxford University Press.

———. 1748/2004. *An enquiry concerning human understanding*. New York: Dover Publications.

———. 1758/1993. *Selected essays*. Oxford: Oxford Univeristy Press.

———. 1777/2006. On the immorality of the soul. In *Essays: Moral, political, and literary*, ed. D. Hume, 597–604. New York: Cosimo Classic.

Kant, I. 1781/1996. *Critique of pure reason*. Indianapolis: Hackett Classics.

Kołakowski, L. 2001. *Metaphysical horror*. Chicago: Chicago University Press.

Krugman, P. 2009, September 6. How did economists get it so wrong? *The New York Times*, 36–43.

Lawson, T. 1997. *Economics and reality*. London: Routledge.

———. 2014. A conception of social ontology. In *Social ontology and modern economics*, ed. S. Pratten, 19–52. Oxon: Routledge.

———. 2015. *Essays on the nature and state of modern economics*. Abingdon: Routledge.

Lenox, J. 2009. *God's undertaker: Has science buried god?* Oxford: Lion Book.

Mackie, J. 1974. *The cement of the universe*. Oxford: Oxford University Press.

Mäki, U. 1992. On the method of isolation in economics. In *Idealization IV: Intelligibility in science*, Poznań studies in the philosophy of the sciences and the humanities 26, ed. Dilworth, 317–351. Amsterdam: Rodopi.

Marshall, A. 1890. *Principles of economics*. 1st ed. London: Macmillan.
———. 1898. Distribution and exchange. *Economic Journal* 8 (29): 37–59.
———. 1920/2013. *Principles of economics*. 8th ed. London: Palgrave Macmillan.
Massey, G.J. 1991. Backdoor analyticity. In *Thought experiments in science and philosophy*, ed. T. Horowitz and G.J. Massey, 285–296. Savage: Roman & Littlefield.
McCloskey, D. 2006. Humility and truth. *Anglican Theological Review* 88: 181–196.
Merton, T. 1956. *Thoughts in solitude*. New York: Farrar, Straus, and Ciroux.
Mill, J.S. 1836/2008. On the definition and method of political economy. In *The philosophy of economics*, ed. D. Hausman, 41–58. Cambridge: Cambridge University Press.
———. 1843. *A system of logic. Ratiocinative and inductive*. London: John W. Parker.
Peirce, C.S. 1877/1982. The fixation of belief. *Popular Science Monthly*. Reprinted in *Pragmatism: The classic writings*, ed. H.S. Thayer, 61–78. Indianapolis: Hackett.
Pizano, D. 2009. *Conversations with great economists*. New York: Jorge Pinto Books.
Polkinghorne, J.C. 1990. A revived natural theology. In *Science and religion. One world: Changing perspectives on reality*, ed. J. Fennema and I. Paul, 87–98. Dordrecht: Kluwer.
Popper, K. 1978. Natural selection and the emergence of mind. *Dialectica* 32 (3/4): 339–355.
Rasmussen, D.C., ed. 2018. *Adam Smith and the death of David Hume: The letter to Strahan and related texts*. New York: Lexington Books.
Robbins, L. 1927. Mr. Hawtrey and the scope of economics. *Economica*: 172–178.
Roberts, R.C., and W.J. Wood. 2003. Humility and epistemic goods. In *Intellectual virtue: Perspectives from ethics and epistemology*, ed. M. de Paul and L. Zagzebski, 257–280. Oxford: Oxford University Press.
Rodrik, D. 2015. *Economics rules: The rights and wrongs of the dismal science*. Oxford: Oxford University Press.
Róna, P. 2018a. Ontology and economics. In *Economic objects and the objects of economics*, ed. P. Róna and L. Zsolnai, 5–8. Cham: Springer.
———. 2018b. Postscript on ontology and economics. In *Economic objects and the objects of economics*, ed. P. Róna and L. Zsolnai, 185–192. Cham: Springer.
Rorty, R. 1995. Is truth the goal of inquiry: Davidson vs. Wright? *The Philosophical Quarterly* 45 (180): 281–300.
Rothschild, E. 2001. *Economic sentiments*. Oxford: Oxford University Press.
Schabas, M. 2008. Hume's monetary thought experiments. *Studies in History and Philosophy of Science* 39: 161–169.
Schumpeter, J. 1954/2006. *History of economic analysis*. London: Routledge.
Senior, N. 1836/1951. *An outline of the science of political economy*. New York: Augustus M. Kelly.
Smith, A. 1759/2013. *The theory of moral sentiments*. London: Empire Books.
———. 1795/1980. *Essays on philosophical subjects, Glasgow edition of the works and correspondence of Adam Smith*. Oxford: Oxford University Press.
St. John Paul, I.I. 1997/2006. *Pielgrzymki do Polski. Przemówienia i homilie [Pilgrimages to Poland: Speeches and Homilies]*. Kraków: Znak.
Walras, L. 1874/1984. *Elements of pure economics: Or the theory of social wealth*. Philadelphia: Orion Editions.
Whitcomb, D., H. Battaly, J. Baehr, and D. Howard-Snyder. 2015. Intellectual humility: Owning our limitations. *Philosophy and Phenomenological Research* 94 (3): 509–539.
Worrall, J. 1989. Why both Popper and Watkins fail to solve the problem of induction. In *Freedom and rationality*, ed. F. d'Agostino and I.C. Jarvie, 257–296. Dordrecht: Kluwer.

Chapter 3
What Is Economics for?

Brendan Hogan

Abstract The methodological foundations of any scientific discipline are shaped by the goals towards which that discipline is aiming. While it is almost universally accepted that the goals of explanation and prediction of natural and non-human phenomena have been met with great success since the scientific revolution, it is almost just as universally accepted that the social sciences have not even come close to achieving these goals. This raises the question addressed in this paper, namely, what is economics, and social science more broadly speaking, for? What is their aim, and how is it similar and dissimilar to that of the natural sciences as we have come to classify them? I take up this question from a pragmatic perspective in this paper, setting economics within the wider context of social inquiry. Specifically, I turn to Hilary Putnam and John Dewey as exemplars of the pragmatic critique of any economics that sees its goals in line with those of the natural sciences, that is, as aiming for explanation and prediction according to governing laws of human behaviour.

The methodological principles that have come to be enumerated as the fundamental starting points of neoclassical economics have been subjected to critique since their beginnings in the Marginal Revolution. The rational agent, the utility maximizing character of their choices, and the methodological individualism that dovetailed so nicely with advancing methods of quantification have all been called into question if not completely refuted from a variety of quarters.[1] These critics come from such areas of intellectual specialization as the philosophy of the social sciences (especially its subbranch the philosophy of economics), disciplines in the humanities outside of the sciences, and even within economics itself. The latter group of critics suffered increasing marginalization as the ascendancy of neoclassical economics married a positivist philosophical underpinning to the scientific pretensions and mathematizing tendencies of economics as a discipline. It is not without some

[1] See, e.g., Anderson (2000).

B. Hogan (✉)
Clinical Associate Professor in Global Liberal Studies, New York University, New York, NY, USA
e-mail: bh72@nyu.edu

P. Róna et al. (eds.), *Words, Objects and Events in Economics*, Virtues and Economics 6, https://doi.org/10.1007/978-3-030-52673-3_3

unintended irony, then, that the practitioners of this science borrowed a distinction from theological discourse and termed mainstream methods 'orthodox' and those outside of the dominant discourse, 'heterodox'.

However, as a now well-told story in philosophical accounts of economics has it, while positivism and specifically the philosophy of science embodied in logical empiricism fell upon hard times in the mid-twentieth century and was bypassed, the mainstream practitioners of economics of at least two schools seemed to ignore entirely the failure of positivism to account for a variety of its own aims. These failures reproduced themselves in neoclassical economics in terms of explanation, the irrational choices agents make, and a consistent failure to generate models which would predict aggregate market behaviour.[2] Behavioural economists took these failures as their starting point in reintroducing irrationality into their understanding of human decisions and now stands as a major contender for explanatory adequacy in providing economics with better scientific grounds for its intellectual project. It also has gained greater political traction in terms of policy. But the question of the relation between values intrinsic to the practice of science and the descriptions generated by that scientific activity itself still remains outside the *organon* of these two major schools of economic thought. This exclusion shows up in two ways. First, both schools basically accept that the end of human activity is the realization of individual preferences. On the neoclassical model, this involves the cost-benefit analysis of an internal algorithm that calculates action based upon given preferences and available information regarding means to satisfy those preferences. Behavioural economics introduces paternalistic interventions at the level of policy prescriptions to overcome the irrationality of individuals in the erroneous choices they make to satisfy their given preferences. Agents exhibit irrationality stemming from a variety of sources including cognitive biases and framing effects that lead to choices which do not maximize utility. Thus, both see action and economically informed policy as a means for preference satisfaction. That is, whether you model practical reason as neoclassical economists do, on what might be characterized as a Humean desire-belief model, or you follow behavioural economists in eliciting the inherent cognitive biases which interfere with our preference satisfaction, you isolate the question of morality from the discipline of economic inquiry. This first shows up in both schools' attempt to provide a value-free and predictive account of economic science.

In this chapter I offer an alternative path for characterizing the practice of economics that is informed by a pragmatic account of social science. On this understanding, as pragmatic philosophers from Charles Sanders Peirce to Hilary Putnam have argued, facts and values are inextricably intertwined in the practice of **all**

[2] The 2008 financial crisis is an example of this failure, a failure so striking as to sponsor a royal commission of leading economists to explain why economists had failed to predict it in the UK. In the US, the testimony of Alan Greenspan before a committee in the House of Representatives has become a touchstone for calls for reform in economics. See in particular his exchange with Rep. Henry Waxman. https://www.c-span.org/video/?c3342718/waxman-greenspan-testimony

science.[3] Additionally, according to the pragmatic view on offer here, sciences on a pragmatic understanding take their cue from the Aristotelian dictum that we should order our method and the goal of our inquiries to the object we are trying to understand. However, somewhat counter-intuitively pragmatism offers a general theory of inquiry as problem solving. It follows from this hybrid of object-specific methods and a fundamental problem-solving aim that if what we are trying to solve are not just problems of physics, for instance, but also the frustration of values and interests of human beings, the inquiry itself will take on a moral character both in terms of vocabulary used to engage the problem, and in terms of the character of the inquiry itself. Thus, according to the pragmatic view in this paper, economics cannot help but be a moral science as it involves issues of labour, distribution, inequality and scarce resources that affect the flourishing of the human species, and now much more broadly, the flourishing of species on the planet. Once economics is redrawn in this way, and the pretensions of economics to be a 'science' modelled on the natural scientific goals of explanation and prediction are reconstructed, the question 'What is economics for? 'might be given a morally and epistemologically robust answer.

In turning to pragmatism, I would like to focus on several features that bring into relief the general orientation pragmatism provides at the epistemological and scientific level. In particular, and what stands in stark contrast to what might be seen as the two main contenders for supremacy in economics mentioned above, pragmatism embeds values in the process of inquiry itself to the point of denying a rigid fact/value dichotomy even with respect to such paradigms of value-neutrality as physics and the rest of the natural sciences. Hilary Putnam was perhaps the most vocal pragmatist in destroying the idea that facts and values are judgments, when expressed as propositions, that can be completely disentangled from each other. It is not only present in his perhaps most famous work, *Reason, Truth, and History*, but it is the topic of an entirely separate book, more than two decades later, *The Collapse of the Fact/Value dichotomy*.[4] These works serve as the foundation for his later work in the philosophy of economics. In *The End of Value Free Economics* with Vivian Walsh. Putnam writes:

> There are facts (using the term as we ordinarily do—not as a term in metaphysical theory, which… is what the logical positivists did) which come into view only through the lenses of an evaluative outlook. 'Virtue terms—terms such as 'brave', 'wise', compassionate', 'resourceful', and their opposites, have indeed figured in philosophical discussions for millennia precisely for this reason.[5]

[3] Putnam has most recently clarified his position on facts and values with respect to ontology, logic, and mathematics as a special case of employing, borrowing a term from Jennifer Case, "optional languages". That these alternative conceptual approaches, what he dubs his 'conceptual pluralism', is a result of not having one true mode of describing the world, and that the choice is based on interests which have an evaluative and rational basis. This pluralism, however, does not have the consequence of relativism or anti-realism. For his late articulation of this thesis, see Putnam (2016).

[4] See Putnam (1980) and Putnam (2004)

[5] Putnam (2003, 396). Repr. in Putnam and Walsh (2012,112)

The consequence of this view is that the attempt in economics to achieve a value-free model of the scientific explanation of human behaviour is fundamentally flawed. Thus, even if economics could imitate the natural sciences as they would like to do, the language involved in describing what is observed or judgments of relevance, to take two elements of scientific inquiry, are value laden. The issue is doubly fraught for the sciences of human behaviour as inquirers employ assumptions with respect to values besides using language that necessarily entangles descriptive and normative elements. Again Putnam:

> The world we inhabit when we describe the world for purposes other than the purposes of physics or molecular biology or some other exact science—certainly the world we inhabit when we describe the world for the purposes the economist is interested in—is not describable in 'value-neutral' terms. Not without throwing away the most significant facts along with the 'value judgments' (Putnam and Walsh 2012, 112).

Besides the issues with thick terms or descriptions that don't allow for disentangling the normative from the factual elements of propositions, another common feature of a variety of pragmatic thinkers with respect to social inquiry is the interests and values made explicit by the fact that something is a problem in the first place. That is to say, it is not just an 'injection' of values via the conceptual and linguistic arguments regarding thick concepts, the entangled character of fact and value in certain terms, but it is also central that problematic situations themselves become available for social scientific inquiry by being constituted in their fabric *as situations* by values.[6] The account that pragmatism, beginning with Peirce, has of inquiry is as a practical activity of moving from the state of doubt to the state of belief. Dewey transformed the doubt-belief matrix while maintaining its general character, into a novel understanding of logic as a theory of inquiry.[7] This extension

[6]The classic statement of this position in the pragmatist literature occurs in John Dewey's *Experience and Nature*. There he writes:

> Or is there an ingredient of truth in ancient metaphysics which may be extracted and reaffirmed? Empirically, the existence of objects of direct grasp, possession, use and enjoyment cannot be denied. Empirically, things are poignant, tragic, beautiful, humorous, settled, disturbed, comfortable, annoying, barren, harsh, consoling, splendid, fearful; are such immediately and in their own right and behalf. If we take advantage of the word esthetic in a wider sense than that of application to the beautiful and ugly, esthetic quality, immediate, final or self-enclosed, indubitably characterizes natural situations as they empirically occur. These traits stand in themselves on precisely the same level as colors, sounds, qualities of contact, taste and smell. Any criterion that finds the latter to be ultimate and "hard" data will, impartially applied, come to the same conclusion about the former. Any quality as such is final; it is at once initial and terminal; just what it is as it exists.

It is beyond the scope of this article to engage the pragmatic resources for this position. It is instructive that after decades of exploring such issues as the role of sense-data in our scientific and metaphysical theories, Putnam increasingly moves towards this position in his writing. See especially his debates with Bernard Williams as evidence of this move. In this shift he relies on Dewey, yes, but in his latest work turns to theories and empirical research in perception to extend his stance. See Dewey (1925, 82) and Putnam (2016).

[7]For the classic statement of this matrix, see "The Fixation of Belief" in Peirce and Buchler (1955).

of Peirce's original formulation is key for both understanding the practical character of scientific activity, but also, crucially, the results of that activity: inquiry is contextually situated and draws on the conceptual resources available for addressing failures of habits at the individual, institutional, and social level. A key element of the contextual character of problematic situations is what Dewey called their qualitative uniqueness. While they share many elements and continuities with respect to previous problem contexts, they are novel. As novel, our patterns of inference, whether deductive or inductive, are not enough to respond in a way that can identify, let alone solve, the problem at hand. Rather, it requires hypothetical inference, or abduction, and experiment.

In addition, on the pragmatic understanding of social science, the practical character of this activity extends beyond the supposed value-neutrality of methodological frameworks to the end towards which economists are aiming. Practices are famously rule-governed activities that have embedded within them goals towards which they are ordered. But what is the goal of economics? Indeed, what is economics for? The model of economics that pragmatism offers takes this question to be one of the primary orienting features of this or any practice that would qualify as scientific inquiry. Briefly, before addressing this goal, it is helpful to highlight two aspects of pragmatism that have deep consequences for how we deal with social problems more broadly: the general character of the pattern of inquiry and the experimental nature of inquiry.

In the pragmatic understanding of inquiry, or science, or how we achieve knowledge about the world (epistemology), pragmatism shares at least one methodological similarity with positivism in at least some of its guises: scientific inquiry has one general pattern. Positivists, in one of its most famous versions, exiled from all meaningful language value propositions as senseless as they had no verification procedure by which they might be tested and secured in the physical sciences.[8] This of course is the opposite of the pragmatic position as values and interests saturate the choices made within problem-solving contexts of all inquiry. The criteria of coherence, for instance, involves evaluation of fit introducing normative judgments into scientific procedures. While pragmatism does share with positivism the theory of a general pattern of inquiry, the practice of doing science and its language is value-laden according to pragmatism. This clashes with positivism, and following upon it, mainstream economics' self-understanding. When it turns to social sciences (especially in the pragmatism of Dewey and Putnam, to point to the main examples in the background of this paper but by no means limited to just these two) this pattern adapts and orders itself to the objects or subjects constituting the problematic context and in addition becomes historically self-conscious with respect to prior methods of inquiry in a cumulative way. Dewey in particular is not sanguine about the difficulties of fulfilling the norms of scientific inquiry with respect to social science, but nonetheless uses it as a normative criterion for parsing what counts as actual social inquiry and what are merely intellectual abstractions based upon *a priori* theoretical commitments.

[8] The two main targets in pragmatist literature taking up this argument are emotivism in ethics and physicalist eliminativism in metaphysics and ontology.

This methodological self-reflection is itself shaped by the social and historical contexts in which it emerges. Put more directly, the very concepts and ideas which guide scientific inquiry as well as what ends up being aritculated as 'fact' are both generated and activated within a context. Beyond the character, then, of our language, concepts, and judgments that betray the infusion of facts and values, the intellectual means at our disposal are operant within a 'background' that is specific to our historical and cultural location. This is a familiar thesis in the philosophy of science that is captured in a different register as the theory-dependence of our data. Putnam's way of putting this within the practical circumscription of our scientific activities is by stating that "science institutes data".[9] That is to say, the facts that are articulated in the language of scientists of course rely on observation, but perception for pragmatists is an interested affair, mediated through the conceptual resources available for making the material of our percepts explicit.[10]

Besides these conceptual features of the pragmatic understanding of inquiry, the general pattern as articulated by Dewey also emphasizes the existential or objective impact inquiry effects. That is, the activity of inquiry reconstitutes problematic situations through the intelligent intervention of humans seeking to solve their problems. Because inquiry is generated out of the doubt of the inquirer in Peirce, or the shattering of our projective habits and anticipations in Dewey, its resolution consists in the reordering of the practical context in which the problem or doubt found its genesis. That is to say, and here we rely on Dewey for its most explicit statement, inquiry involves resolving problems that emerge from breakdowns of our practices in such a way as to reconstruct and restore the environment:

> Inquiry is the controlled or directed transformation of an indeterminate situation into one that is so determinate in its constituent distinctions and relations as to convert the elements of the original situation into a unified whole.[11]

I enlist this oft-quoted definition for two reasons. The first is that it raises a whole host of issues as to how to understand the purposes of the sciences themselves, and these issues only become more complicated when dealing with human subjects. Far from limiting the purpose of science as creating a verified system of propositions that describe the natural world in a value-neutral way, pragmatic inquiry is involved in actively constituting the problematic situation with respect to the conceptual resources at hand and then engaging in experimental activity that attempts to solve the problem. In fact, it is in discussing social inquiry as a special instance of inquiry that Dewey takes pains to reiterate the world-changing or interventionist character of natural scientific inquiry.

---

[9] Rorty et al. (2004). This conversation between James Conant, Hilary Putnam, and Richard Rorty provides an accessible and illuminating exploration of these issues, along with Putnam's disagreements with Rorty.

[10] I state the matter this way for sake of expediency. There is a much longer pragmatic story to tell regarding the relation of concepts, percepts, and stimuli.

[11] See Dewey (1938, 109).

3.1 The Philosophy of Social Science and Pragmatism

Here I would like to retreat to a higher level of abstraction, but one that allows for an understanding of how crucial the question of this reflection is to the practice of economics itself from a pragmatic perspective. When we place economics as a discipline grouping it under the social sciences, we can help illuminate an answer as to what it is for by unearthing some of the general features of inquiry in the social sciences, tout court. The debate over what the social sciences are for can be captured very generally by referring to the three ends of explanation, understanding, and emancipation.[12] That is, what social scientists generally consider the goal of a social scientific activity have tended to sort out along three lines:

1) The social sciences aim at results akin to the explanations based on cause -effect covering law models as in the natural sciences, or some lesser version of inductive correlation.
2) The social sciences are geared towards mutual understanding of the meaningful and expressive character of an animal that is essentially self-defining and self-articulating through actions not reducible to a framework available to an independent observer.
3) The social sciences are part of an overall praxis whose ultimate goal is emancipation from the irreducible dimensions of power that have thus far stratified societies in ways that arbitrarily oppress, dominate, and diminish human creativity through the systematic distortion of the intentions of actors.

The naturalist, interpretive and critical schools of social science then, are part of the conceptual inheritance of social science when it comes to the intellectual and practical task of addressing economic problems. Here we turn to the pluralism of pragmatism, and in particular to John Dewey's philosophy of social science, to address the purpose of economics, vis-*a-vis* the philosophy of social science.

In the face of such different and at first face incompatible ends for economics as a special example of social inquiry, the pragmatic approach famously employs what might be termed a Hegelian strategy in a metaphysically naturalist vein. That is, rather than select one of the methodological platforms for the purpose of social science, pragmatism absorbs each in a pattern of inquiry that draws out the lessons for practical coping with problems that each method was historically inspired to address. Each of these schools helps as a stage in articulating and addressing the problem at hand. In addition, because problematic situations are unique and composed of variegated elements in differing intensities of influence, no one methodological platform has priority of others in an *a priori* fashion. In some instances what we observe using utility maximization models of human choice is correct. In others, it would be misleading to impose these as explanations of action.[13] This makes the employment of

[12] See the classic statement of these options see Fay and Moon (1977).

[13] Clifford Geertz's (1973) "Deep Play: Notes on the Balinese Cockfight" remains an instructive touchstone for the dangers of imposing rational choice models on social phenomena.

method a practical affair requiring an experimental and fallibilistic attitude: Are there causal elements to human behaviour that need be elicited to help explain the problematic situation? Am I really understanding what is going on in this action context without engaging the participants in ways such that I empathetically reconstruct their intentions so as to give meaningful sense to their actions? Are there forces at work psychologically distorting practices and intentional frameworks for the purposes of ideological manipulation? These are all questions that will help illuminate human action in general, and what has come to be defined as economic action over time. This translation or, if you prefer, sublation of different paradigms of social science and, by extension, economic schools is not limited to pragmatic philosophy broadly to the social questions of this sort. Rather, pragmatism's overall strategy is to see each of the alternatives developed in the past with regard to ethics, epistemology and metaphysics as methodological frames by which to assist in reconstructing a problematic situation in the present into a resolved whole.[14]

The particularly pragmatic contribution in social science, as a potential fourth element to social inquiry, is its experimental character. Because Dewey's philosophy of social science remains the most detailed working out of the pragmatic turn in philosophy with respect social inquiry, I will rely on his discussion of it in his 1938 *Logic: the theory of inquiry*. There Dewey marks out in definite terms what the consequences for the pragmatic reconstruction of the history of philosophical inquiry are for the social sciences in the penultimate chapter of that work. He states that social inquiry is especially vulnerable to the positivist doctrine of the strict separation of facts and values and thus the independence of social science from social practice:

> …the idea commonly prevails that such inquiry is genuinely scientific only as it deliberately and systematically abstains from all concern with matters of social practice. The special lesson which the logic of the methods of physical inquiry has to teach to social inquiry is, accordingly, that social inquiry, *as inquiry*, involves the necessity of operations which existentially modify actual conditions that, as they exist, are the occasions of genuine inquiry and that provide its subject-matter. For, as we have seen, this lesson is the logical import of the experimental method.[15]

That is, for pragmatic social science, social action for the sake of addressing a problem articulated in the different methodological processes of social inquiry is endogenous to social inquiry itself.

---

[14] I have referred to these questions as informing different methodological moments of social inquiry in Hogan and Marcelle (2017). This pluralist strategy has some current practitioners in contemporary economics. Ha-Joon Chang, for instance, has famously been arguing both in publications and making popular videos that at once reorient the abstractions and expert level of economics and also challenges the orthodoxy that, while on its back heels intellectually speaking, still maintains sociological and political dominance. Chang in fact lists 11 different schools of economics. Also resonant with pragmatism is Chang's insistence that the experience of an individual is a sufficient starting point for making informed judgments about economic systems and their outcomes. See Ha Joon Chang (2011) and RSA- Animate: Economics for Everyone- a cognitive whiteboard animation. https://www.youtube.com/watch?v=E9EzXHVYCIl

[15] Dewey (1938, 486)

Social inquiry is about solving problems, as is all inquiry, and solving the problems of human beings requires social cooperative action. A problem is not solved by a journal article or a monograph, no matter how brilliant. It is here that the larger goals of community life are inscribed within each of the social sciences. That is, each social science as it has developed is actually not even scientific, according to pragmatism, unless it involves cooperative, coordinated action on the part of the individuals who live within the problematic context to solve the problem:

> That which is observed, no matter how carefully and no matter how accurate the record, is capable of being *understood* only in terms of projected consequences of activities. In fine, problems with which inquiry into social subject-matter is concerned must, if they satisfy the conditions of scientific method, (1) grow out of actual social tensions, needs, "troubles"; (2) have their subject-matter determined by the conditions that are material means of bringing about a unified situation, and (3) be related to some hypothesis, which is a plan and policy for existential resolution of the conflicting social situation.[16]

Our only path for confirming our social scientific hypotheses, however, is through coordinated action. The epistemological requirement of getting the problem right in the first place includes the public or practical verification of the problem through channels of communication that take up into the problem formation process the perspectives of the subjects themselves.[17]

This aspect of social inquiry widens the practical requirements of the social sciences into forming a public that becomes a constitutive feature of problem formation. In contradistinction to appeals to technocratic management of political policies and legislation, the problems of political, economic, and social life are here understood as products of the articulation and communication of those individuals who are experiencing the problem themselves. Because problematic situations do not exhibit their constitutive features in ways that are easily legible to an observer, the demand to figure out what the problem in the first instance is paramount. But rather than these problems being of the nature of different chemical compounds and their reaction to each other, or the motion of planetary bodies, social problems involve agents who disagree, come into conflict, resort to violence, and have widely varying interpretations of the cause and effect processes and powers governing their life-chances. If social scientists are going to have a chance at getting the problem right, they themselves must suspend their *a priori* predilections to define social problems according to a vocabulary and methodology that worked in a prior historical instance. This is not to elide what I earlier referred to as the contextual and historically specific character of our conceptual employments. It is to recognize the fallible and situated character of our conceptual projections in light of a novel situation. The qualitative uniqueness, however, that is the mark of problematic situations prevents inquirers from apprehending in a cognitively thorough manner environments that exhibit these various unique characteristics, spontaneous energies, and novel constellations of forces at work. In short, how we come to a robust description of the

[16] Dewey (1938, 493)

[17] On the concept of practical verification, see Bohman (2003).

actual problem at hand is through accessing those problems such that any type of possible hypothetical solution addresses the problems as experienced by the subjects who suffer them:

> The connection of social inquiry, as to social data and as to conceptual generalizations, with practice is intrinsic not external. Any problem of scientific inquiry that does not grow out of actual (or "practical") social conditions is factitious; it is arbitrarily set by the inquirer instead of being objectively produced and controlled. All the techniques of observation employed in the advanced sciences may be conformed to, including the use of the best statistical methods to calculate probable errors, etc., and yet the material ascertained be scientifically "dead," i.e., irrelevant to a genuine issue, so that concern with it is hardly more than a form of intellectual busy work. That which is observed, no matter how carefully and no matter how accurate the record, is capable of being *understood* only in terms of projected consequences of activities.[18]

One of the constraints, then, on this understanding of social science is that problem formation itself relies upon the articulation of the frustration of interests and harms experienced by individuals such that our understanding of the problem can be as epistemologically robust as possible. It is of course possible, that subjects themselves articulate their perspectives saturated in misinformation, or prejudice, or ignorance. However, the process of social inquiry involves the public in a self-reflective way precisely to discover and address these deficits, and to see how these errors themselves contribute to the problem at hand.

In this sense the critique of the one-dimensional character of the dominant schools of economics is a moral one. Economics is for solving economic problems, and the path to solving these problems, as pragmatism understands it, requires a wide array of methods and approaches to even begin to get the problems of economy right in the first place. The problems are defined in conjunction with the articulation and action of the subjects in the problematic context. Though it has now become increasingly popular to criticize the mainstream and policy powering branches of economics, Dewey articulated his critique and view of social science in the 1930s culminating in 1938, and it is not coincidental that the context within which Dewey was writing was the Depression and the concomitant rise of fascism and solidification of Soviet communism. In developing a pragmatic understanding of social inquiry and economics as a special instance of social inquiry at this historical conjuncture, Dewey was trenchant in his critique of all forms of political doctrine founded upon an understanding of human action and human society that reduced the *explanandum* of history to the *explanans* of economic agency.

While this historical comment is illustrative, it serves a conceptual as well as exemplary purpose. For if pragmatic inquiry, as inquiry, is to get a problem right, the unsparing criticism of all forms of economic methodology and scientific practice that hypostatize principles of inquiry into *a priori* certainties that control data selection and formulation is called for. Thus, while Dewey clearly articulated a devastating critique of Soviet economism in his *Freedom and Culture* (1939) and referred to what had become of Marx's thought in the service of the Russian

[18] Dewey (1938,492)

revolution as "totalitarian economics", he was no less penetrating in his critique of the political and economic doctrines concomitant with liberal political economies in the industrial world.[19]

To close, I have focused on the general features of those aspects of pragmatism that solidify its approach to economics as a moral science. I have turned to the fact/value distinction, the philosophy of social science, and the experimental character of inquiry pragmatic philosophy of social science calls for in order to demonstrate this link. The precise way in which this becomes a moral issue is through the necessity of enlisting the values and perspectives of the subjects themselves in constituting the problems social inquiry deals with under the rubric of 'economics'. These conceptual and practical elements are not the only ways in which pragmatic models of social inquiry are relevant to a project of reconstructing economics to make plain its moral purpose. Specifically, the model of the agents and the availability of preferences for rational evaluation in choice situations is another rich path Putnam and others have taken in combining ethics and economics. This is yet another way in which two of the dominant schools of economics wielding power in policy and intellectual culture, neoclassical and behavioural economics, fall short from a moral perspective in the practice of economics.

References

Anderson, E. 2000. Beyond homo economicus: New developments in theories of social norms. *Philosophy and Public Affairs* 29 (2): 170–200.

Bohman, J. 2003. Critical theory as practical knowledge: Participants, observers, and critics. In *The Blackwell guide to the philosophy of the social sciences*, ed. Stephen Turner and Paul Roth. London: Blackwell Press.

Chang, Ha Joon. 2011. *23 things they didn't tell you about capitalism*. New York: Penguin. Reprint edition.

Dewey, J. 1925/1985. *The later works of John Dewey, 1925–1953. Volume 1: Experience and nature*. Carbondale, IL: Southern Illinois University Press.

———. 1930. *Individualism old and new*. New York: Minton, Balch & Company.

———. 1935. *Liberalism and social action*. New York: G.P. Putnam.

———. 1938/1985. *The later works of John Dewey, 1925–1953. Vol. 12: Logic: The theory of inquiry*. Carbondale, IL: Southern Illinois University Press.

———. 1939. *Freedom and culture*. New York: G.P. Putnam.

Fay, B., and J.D. Moon. 1977. What would an adequate philosophy of social science look like? *Philosophy of the Social Sciences* 7 (3): 209–227.

Geertz, C. 1973. Deep play: Notes on the Balineses cockfighting. In *The interpretation of cultures: Selected essays*. New York: Basic Books.

[19] For Dewey's critique of liberalism and the varieties of laissez-faire economic methodologies informing public policy in his time, see Dewey (1935) and Dewey (1930). It is beyond the scope of this paper to articulate in detail Dewey's criticisms. However, they offer a clear example of how the pragmatic logic of social inquiry handles the basic assumptions of a variety of elements in the neoclassical approach to economics as well as liberalism's conceptual foundations more generally.

Hogan, B., and L. Marcelle. 2017. The complementarity of means and ends: Putnam, pragmatism, and the problem of economic rationality. *Graduate Faculty of Philosophy Journal* 38 (2).
Peirce, C., and J. Buchler. 1955. *Philosophical writings of Peirce*. New York: Dover Publications.
Putnam, H. 1980. *Reason, truth, and history*. Cambridge, MA: Harvard Press.
———. 2003. For ethics and economics without the dichotomies. *Review of Political Economy* 15 (3).
———. 2004. *The collapse of the fact/value dichotomy and other essays*. Cambridge, MA: Harvard University.
———. 2016. *Naturalism, realism and normativity*. Cambridge, MA: Harvard University Press.
Putnam, H., and V. Walsh. 2012. *The end of value-free economics*. London: Routledge.
Rorty, R., H. Putnam, J. Conant, and G. Helfrich. 2004. What is pragmatism? *Think* 3 (8).

Chapter 4
Should Economics Make a Pragmatic Turn? John Dewey, Karl Polanyi, and Critique of Economic Naturalism

Maciej Kassner

Abstract This article draws upon John Dewey's pragmatic philosophy to offer a critique of naturalistic assumptions underpinning modern economic thought. More specifically, mainstream economics is criticized for treating its subject-matter as human-independent, maintaining untenable means-ends dualism, and displaying a strong tendency to talk about capitalism and markets as if they were quasi-natural phenomena. Pragmatic arguments are developed in dialogue with the ideas derived from Karl Polanyi, Gunnar Myrdal, and contemporary philosophy of science. Throughout the article, institutionalism is singled out as an approach to economic inquiry that is compatible with central insights derived from Dewey's version of pragmatism. The article suggests that pragmatism together with constructivism and hermeneutics provides a more appropriate philosophical framework for institutional economics than other approaches, including critical realism.

4.1 Introduction

In this essay, I will argue that the philosophy of economics – and economic thinking more broadly – should make a pragmatic turn. Yet, pragmatism is not a uniform tradition and critics vary greatly in their assessment of various aspects of pragmatic legacy. Here I wish to concentrate on a version of pragmatism developed by John Dewey since it seems to be most relevant for economics and social sciences in general. My goal in this essay is to appropriate certain insights taken from Dewey's philosophy in the critique of contemporary economic discourse. Following Dewey, I wish to argue that economics, despite its loudly proclaimed scientific aspirations, is still held captive by unexamined assumptions derived from various versions of

This research project was supported by funding form National Science Centre, Poland (No. 2016/20/S/HS5/00556).

M. Kassner (✉)
Faculty of Political Science and International Studies, University of Warsaw, Toruń, Poland

P. Róna et al. (eds.), *Words, Objects and Events in Economics*, Virtues and Economics 6, https://doi.org/10.1007/978-3-030-52673-3_4

philosophical naturalism.[1] In *Logic: The Theory of Inquiry* Dewey argued that "the net consequences of classical economics was reinstatement of the older conception of natural laws by means of a reinterpretation of their content" (Dewey 1938, 498). This remark was directed against the old natural law philosophy, which inspired the classical economics of Adam Smith and his followers. Similar critique, I would argue, can be raised against contemporary neoclassical theory and its allied approaches, which based their scientific aspirations on a particular understanding of physics and other natural sciences.[2]

More specifically, strong naturalistic influence can be detected in at least three areas of contemporary economic thinking. First, the dominant self-understanding of the economic profession assumes that the economy can be studied in the same way we study nature. This implies that the subject-matter of economics is treated as ontologically independent from human intentions and purposes. Second, neoclassical economics is based on a very problematic view of human nature and rationality, which treats means-ends dualism as a universal feature of rational behaviour. Thirdly, naturalistic undertones can be discerned in various conceptualizations of capitalism or the market economy which treat them as quasi-natural phenomena driven by their own internal laws. When applied to the subject matter of economics Deweyan pragmatism can be seen as a form of philosophical therapy. By exposing hidden philosophical presuppositions of contemporary economic theory pragmatism can loosen its grip over the economic profession and make economists and other social scientists more willing to examine alternatives to contemporary orthodoxy.

It should be clear from the outset that the argument offered in this essay is not neutral between various research traditions within contemporary social sciences and rival economic ideologies. On the contrary, it takes a partisan stance and champions the institutional approach to the economy as represented by original institutional economics and various research traditions within political science, sociology, and anthropology. More particularly, it tries to combine pragmatic arguments with lessons derived from Karl Polanyi and research tradition inspired by his work. Interestingly enough, Deweyan pragmatism and Polanyian institutional approach to political economy are not only compatible on a theoretical level but also embedded in a very similar social and political outlook. John Dewey and Karl Polanyi were

[1] For students of Dewey's thought this assertion may be surprising. Was not Dewey himself a proponent of naturalism both in general and with regard to social inquiry? The answer to this question is of course positive. But Dewey's version of naturalism had nothing to do with the naturalistic ambition, which drives contemporary economics. Suffice to say that for Dewey the presence of practical purpose and value judgments are two universal traits of all types of inquiry.

[2] Although orthodox economics can be easily identified, providing an exact definition of it may be difficult. In this paper such terms as mainstream or orthodox economics will refer to various approaches that are basically consistent with the central features of neoclassical analysis (i. e. mathematical formalism, rationality defined as utility maximization, and equilibrium analysis), even if they modify some of its assumptions. On definitional issues see further Arnsperger and Yannis (2008); Dequech (2007) and Lawson (2006).

almost contemporaries.[3] Both thinkers offered largely parallel critiques of nineteenth-century *laissez-faire* ideology and naturalistic philosophical assumptions behind it. More importantly, they both tried to avoid false alternatives represented by orthodox liberalism and orthodox Marxism. In order to overcome the ideological limitations of his time, Polanyi sought to examine the place of markets in different economic systems which existed in the past.[4] His radical institutionalism, which emphasizes the fact that markets are always socially constructed, resonates well with Deweyan anti-essentialism. Taken together these approaches suggest a novel and compelling way to think about the role of markets and institutions in a contemporary society.

We can now state our initial thesis in a more precise form. I believe that taking the pragmatic turn would make economics less naturalistic, more institutionally oriented and more radical in its social outlook. I will try to make this argument in the following steps. Firstly, I will examine the present state of economic discourse and try to show that some of the shortcomings of modern economics stem from its unconscious reliance on very problematic philosophical assumptions. Secondly, I will contrast Deweyan social psychology with simplistic utilitarianism, which dominates contemporary economic thought. Finally, I will suggest some lessons that we can learn from Dewey and Polanyi regarding the place of markets in a just social order. In the concluding section of the paper, I ask what, if anything, economics can gain from Deweyan pragmatism and Karl Polanyi's social thought.

4.2 Our Problematic Situation: Mainstream Economics and the Legacy of Naturalism

John Dewey teaches us that all thinking has its beginning in some deeply felt dilemma or trouble.[5] Arguably, Dewey's thesis has been vindicated by the rise of philosophical reflection regarding economic subject-matter. Indeed, the growing popularity of such a relatively young subfield as philosophy of economics indicates that professional economics is beset by deeply felt scepticism concerning the validity of its assumptions, models and standard methods of conducting research. There is a widespread conviction that not all is well with the way economics is thought and practiced today. Economics, it is often said, is a science of choices that people make under conditions of scarcity.[6] If so, economic methods and concepts are largely irrelevant for understanding more complex social situations such as, among other things, the present state of economic science. Ordering preferences and listing pos-

[3] For biographical information consult Westbrook (1991) and Dale (2016).

[4] See Polanyi (1977).

[5] See Dewey (1938, 109).

[6] For the classic formulation of the scarcity definition of economics see Robbins (1932, 15).

sible choices is not a suitable panacea for a deep crisis of professional self-confidence. We need to engage in a more interpretative inquiry and ask what kind of cultural tensions reveal themselves in contemporary philosophical debates about the present state of economic science.

The best way to start the inquiry is to take a closer look at its scientific ambitions. Philosophical self-understanding of mainstream economics, to the extent that it has any, is defined by naturalistic outlook. By naturalism we mean here the desire to imitate the most successful of the natural sciences, most particularly physics. As Philip Mirowski (1989) has argued in his book *More Heat Than Light* modern neo-classical economics has started out as an attempt to use concepts derived from nineteenth-century energy physics to the study of economic subject-matter. From that time on, recurrent physics envy is one of the most enduring characteristics of mainstream economic thinking. In the early twentieth century, self-understanding of naturalism was transformed under the influence of new philosophy of science represented by logical positivism and the teachings of Karl Popper. Influence of positivism is evident in the writings of such luminaries as Paul Samuelson and Lionel Robbins (Macpherson 1983, 97). Arguably, the naturalistic orientation of contemporary mainstream economics can be characterized as an attempt to imitate a particular image of natural science, which was created by the positivists and the Popperians after World War II. In his book devoted to the critical examination of the idea of social science in the philosophy of Charles Taylor and Alasdair MacIntyre, Jason Blakely noted that neoclassical economics is a "standard-bearer for naturalism in the social science today".[7] Economics is also a major source of naturalistic influence for other social disciplines, most notably political science. As many authors observed, this naturalistic orientation is revealed in a predominance of mathematical formalism, which bears some superficial resemblance to techniques used in natural sciences.[8] Even more importantly, naturalism brings with itself certain problematic philosophical assumptions about the character of the subject matter of economics.

At the heart of naturalism lies the idea that the economy resembles the natural world at least in one essential regard. Both economic and natural phenomena are thought to be subject to laws and regularities that are ontologically independent from human will and intentionality. From the point of view of Dewey's pragmatism, modern economics is a special case of the spectator theory of knowledge. According to this fallacious view, knowing is conceived as radically separated from doing. As a consequence, the subject matter of knowledge is conceptualized as ready-made and complete in itself. In *The Quest for Certainty*, Dewey offered the following explanation of the consequences of the spectator theory of knowledge with special reference to economic thinking:

> The doctrine that nature is inherently rational was a costly one. It entailed the idea that reason in man is an outside spectator of a rationality already complete in itself. It deprived

[7] See Blakely (2016, 17).

[8] On the contested issue of mathematical formalism in economics consult Mirowski (1986).

> reason in man of an active and creative office; its business was simply to copy, to re-present symbolically, to view a given rational structure. (…) The doctrine was both an effect of the traditional separation between knowledge and action and a factor in perpetuating it. (…) Its paralysing effect on human action is seen in the part it played in the eighteenth and nineteenth century in the theory of natural laws in human affairs, in social matters. These natural laws were supposed to be inherently fixed; a science of social phenomena and relations was equivalent to discovery of them. Once discovered, nothing remains for man but to conform to them; they were to rule his conduct as physical laws govern physical phenomena. They were sole standard of conduct in economic affairs; the laws of economics are "natural" laws of all political action; other so called laws were artificial, man-made contrivances in contrast with the normative regulations of nature itself. *Laissez faire* was the logical conclusion. For organized society to attempt to regulate the course of economic affairs, to bring them into service of humanly conceived ends, was a harmful interference (Dewey 1929, 169).

While contemporary economics cannot be said to be uniformly pro-market in its ideological orientation, it nevertheless retains essentially the same notion of fixed subject-matter that was criticized by Dewey. The essence of this view is to treat economics as a discipline concerned solely with a value-free study of pre-existing economic reality.

The naturalistic view of the relationship between economics and its subject-matter rests on very problematic epistemological and ontological assumptions. On the epistemological level, it is questionable whether we really can know the economy – or, for that matter, anything else – without taking into consideration practical purposes which our knowledge should serve. The thesis that knowing and doing are always connected in the process of inquiry is one of the distinguishing traits of pragmatism. The essence of pragmatic position was captured well by Clarence Irving Lewis who said that for Dewey 'meaning and action are necessarily connected" (Lewis 1939, 572). As Lewis explains Dewey's logic contains

> …distinctive conception, incompatible with most views, that the cognitive or meaning situation does not admit of bifurcation into an activity of the knower and a preformed object which is contemplated; that knowing or meaning is integral with other activities which shape the objects to which they are addressed; that meanings themselves serve to frame the situations of action into which they enter, and exercise an operational force upon what they serve to formulate. It is implied that an idea or a meaning, apart from some possible action and the reality in which it should eventuate, is a fictitious entity not found in human thinking. And conversely, it is implied that the objects of knowledge, without reference to meanings and the actions to which they may lead, are equally fictitious (1939, 572).

We can glimpse from this quotation how radical the pragmatic theory of knowledge really is. It seeks to undermine basic dualisms of modern philosophy including those between subject and object, theory and practice, and knowing and the known. What we get instead is a picture of inquiry as a form of practical activity whose ultimate aim is to reconstruct its subject matter according to some purpose or plan. Thus, the goal of inquiry is not only to understand the world but also to change it or, said more precisely, theoretical understanding of the world would be impossible without some sort of practical purpose which guides our interpretation of a given problematic situation.

The interconnection between cognitive situation, meaning, and practical purpose is evident in Dewey's conception of the mutual interdependence of facts and values.[9] According to Dewey, judgments about facts and judgments about values are closely linked in the process of inquiry. Every inquiry takes place in the context of some specific problematic situation. Selection of data and facts is guided by implicit or explicit determination of their relevance to the issue at hand. Without reference to some problem we do not have facts but only a mass of meaningless information. But to say that facts are relevant to a given case is to make a value judgment since relevance is obviously an evaluative term. More generally, the way we organize our facts depends to a large extent on the value premises of our theories. Strikingly similar point was made by Gunnar Myrdal with regard to economic inquiry. According to Myrdal "all science, at any rate all social science, is «practical», even in its purely casual theory, because it must choose one from amongst infinite number of possible ways of collecting and ordering its infinitely large mass of empirical data" (1958, 209). Thus, every principle of selection contains some overt or hidden reference to practical purpose. Historically speaking, economic theories emerged as an attempt to meet the demands of some "concrete and particular political questions" (1958, 210). Thus, for Myrdal "practical political economy", which is defined by its adherence to particular aims and purposes, "logically precedes any causal explanation" (1958, 210). Thus, any strict distinction between theory and practice, or positive and normative economics is untenable.

Problematic situations, the main subject of social inquiry, are practical in yet another sense. As John Dewey noted, problems investigated by social sciences typically "grow out of actual tensions, needs, «troubles»" (Dewey 1938, 493). Charles Wright Mills, who followed Dewey in this regard, argued that a social issue "involves a crisis in institutional arrangements, and often, too, it involves what Marxists call «contradictions» or «antagonisms»" (2000, 9). In the context of particular social conflicts, facts can be divided into factors that enable or inhibits the realization of certain desired goals (Dewey 1938, 493). Following this further, we can note that the perception of social problems is strongly influenced by the perspective of particular social actors. Businesspeople, trade unionists, environmental activists, and feminists can have different ideas about what counts as an economic problem and which data are particularly relevant. Taking this into consideration, James Bohman (2002) argues for multiperspectival social science, which includes points of view of all relevant actors. Moreover, competing ideologies and social philosophies can be interpreted as different ways of making sense of problematic situations. They offer divergent hypothesis regarding the nature of social problems that should be given a free play in the course of economic investigations. Thus, it is vital for an economic inquiry to be open to different theoretical, ideological and methodological perspectives. Interpreting the world through the lenses of a single

[9] For Dewey's view see in particular his essay *The Logic of Judgments of Practice* (1915) MW 8, 14–97. For perceptive commentary consult Bernstein (1999, 213–219).

theory is likely to lead to a strong bias, regardless of how many Nobel prize laureates in economics may be associated with that particular theory.[10]

In human sciences, facts and values are connected not only epistemologically but also ontologically. Dewey made this point in *The Public and Its Problems* when he observed that "political facts are not outside human desire and judgment. Change men's estimate of the value of existing political agencies and forms; and the latter change more or less."[11] What seems obvious with regard to political institutions, such as constitutions and electoral laws, also remains true for economic institutions like money, credit, and banking. In contemporary philosophy similar conception of the subject-matter of social sciences has been put forward by the proponents of constructivism and hermeneutics. Following John Searle, we can say that the subject matter of social sciences, economics included, is epistemologically objective but ontologically subjective (Searle 2005, 3–5). To say, that the economy is epistemologically objective is to maintain that more or less warranted assertions can be made about it. To say that the economy is ontologically subjective is to assert that it is constituted by institutions, which are collectively created with the use of language. In other words, economic objects owe their very existence to human intentionality and judgment as well as moral purpose embodied in them (Róna 2017, 3–9). For instance, such business artefacts as money and banking are unthinkable without rules and shared meanings, which are associated with them. Hence, interpretations are a constitutive part of economic reality.[12] Strangely enough, the fact that economic reality is laden with intersubjective meanings and purposes is systematically ignored by mainstream economic theory. Fear of allegedly subjective interpretations is evident in the tendency of mainstream economics to treat preferences as a given property of autonomous individuals.[13] As a result, culturally specific and socially constructed character of human wants and preferences is placed outside of the scope of orthodox economic inquiry.

In this section, I have argued against naturalistic self-understanding of economic profession, which states that the goal of economics as a positive science is to produce value-free explanatory theories about the nature and workings of the economy. I have found this self-understanding inadequate on both epistemological and ontological grounds. From an epistemological perspective, a positive conception of economic inquiry is neither possible nor desirable. It is impossible since we cannot have any scientific discourse without judgments related to such values as significance, simplicity, fruitfulness, adequacy, and so on.[14] It is undesirable since the attempt to construct such science makes it more difficult to use the insights of economic inquiry for the purpose of social and economic reconstruction. From the

---

[10] For recent arguments in favour of a greater pluralism in economics see Fullbrook (2008, 13–25).

[11] See Dewey (1927, 240). For an elaboration of this point with respect to economics see Dewey (1947, 224-238).

[12] For a classic statement of interpretative conception of social inquiry based on the view that social reality is constituted by shared meanings see Taylor (1971).

[13] See Macpherson (1983).

[14] For a further elaboration of pragmatic critique of fact/value dichotomy see Putnam (2002).

ontological perspective, the main shortcoming of mainstream economics lies in the neglect of the role of ideas, interpretations, and meanings in the constitution of economic subject matter. Ultimately, economic reality consists of historically specific practices and institutions, which cannot be adequately conceptualized without taking into account the category of a shared culture.

4.3 Overcoming the Limits of Rational Choice: Pragmatism and Institutionalism

In addition to the conception of economic subject-matter shaped by naturalism, neoclassical economics also contains a view of human being as *homo economicus*, devoted single-mindedly to the maximization of utility. Modern economics assumes universal human nature and universal rationality, which consists of the capacity to choose the most efficient means to a given set of aims. For the followers of mainstream economics, all human beings from diverse cultures and different historical epochs are essentially alike, at least with regard to the type of rational calculations that guide or should guide their conduct. Hence, categories developed by neoclassical economics are believed to be rooted in the very nature of human beings. But a closer examination of the philosophical assumptions underlying economic theories of human nature and rational action reveal that they are built on very shaky foundations. From the pragmatic perspective at least two elements of the received view are extremely problematic. Firstly, the idea of universal human nature breaks down when confronted with a variety of culturally specific habits and institutions. Secondly, the rigid dualism of means and ends seems to be untenable in the light of a pragmatic theory of action which revolves around the idea that means and ends are "two names for the same reality" (Dewey 1922, 36).[15] That is to say, the distinction between means and ends is not absolute but contextual. Those arguments, taken together, undermine the behavioural micro-foundations of modern economics.

In order to expose questionable behavioural assumptions behind much of present-day economic thinking, I shall compare social psychology of habits developed by Dewey in *Human Nature and Conduct* with individualistic psychology of utility maximization that dominates mainstream economics. Dewey offers a broad definition of a habit as "that kind of human activity which is influenced by prior activity and in that sense acquired; which contains within itself a certain ordering or systematization of minor elements of action; which is projective, dynamic in quality, ready for overt manifestation; and which is operative in some subdued subordinate form even when not obviously dominating activity."[16] It is important to understand that for Dewey habits are functions of social surroundings. They are not inborn but acquired through the process of socialization and participation in culture. They

[15] For a theory of creative action build on Dewey's social psychology see Joas (1996).

[16] See Dewey (1922, 31).

denote typical modes of response to standardized social situations. In this respect, they are similar in meaning to the concept of institutions. Affinity of the two terms was perceptively noted by Wesley Mitchell, a leading institutional economist and Dewey's student at the University of Chicago. As Mitchell observed, the term "institutions'" as conceptualized in institutional economics is merely a convenient name for "the more important among the widely prevalent, highly standardized social habits" (1950b, 373).[17] However, rootedness of economic action in the of shared habits and institutions is systematically ignored by neoclassical economics and akin approaches. Indeed, as Geoffrey Hodgson observed, one of the major limitations of mainstream economics is that it is institutionally blind.[18] Hence, the integrity of the behavioural assumption behind mainstream economics can be preserved only at the price of ignoring or downplaying the importance of the institutional context of the economic inquiry.

Another problematic aspect of mainstream economic theory lies in its attempt to introduce exact mathematical calculus into the study of human conduct. As Dewey noted, future pleasures and pains are "subject to incalculable accident" (1922, 38). Similar arguments apply also to the notion of marginal utility, which is nothing other than an imaginary unit of preference satisfaction. Clearly, calculations made in such terms can bring only a distant semblance of exactness and rigor. On the contrary, such categories as habits and institutions cannot be easily expressed in mathematical formulas. Nevertheless, they have a considerable advantage from an empirical point of view. To put it simply, habits and institutions are revealed in specific patterns of behaviour, which can be recorded and studied with the help of appropriate quantitative and qualitative techniques. Norms guiding human conduct can be studied through participant observation, in-depth interviews, discourse analysis and other methods derived from anthropology and qualitative sociology. However, institutionalism does not need to shy away from the study of macro-phenomena with the use of quantitative methods. Indeed, as Wesley Mitchell, an institutional economist and one of the founders of modern econometrics observed, "the quantitative workers will have a special predilection for institutional problems because institutions standardize behaviour and thereby facilitate statistical procedure" (1950a, 30). Mitchell's approach, although biased in favour of statistical methods, is poles apart from the use of econometric techniques prevalent in neoclassical economics, where "the business of statistician is merely to verify conclusions established by deduction" (1950a, 33). To conclude, from a pragmatic perspective, studying patterns of behaviour observed in the fieldwork or encapsulated in time series of economic data, is eminently more enlightening than building formal models based on aprioristic assumptions.

Equally serious difficulties emerge when we look more closely at means-ends dichotomy, which is assumed by most economics textbooks. Mainstream economic theory treats economic rationality as a matter of finding the most efficient means to

[17] Mitchell was Dewey's student at the University of Chicago.

[18] See Hodgson (2002).

given ends. But such account is ridden with serious difficulties. To start with, our ability to act rationally is predicated on the possibility of having adequate knowledge about future states of the world. Unfortunately, social sciences, including economics, failed to demonstrate their capability of yielding such knowledge. Following Mary Hesse, we may say that natural sciences are instrumentally progressive (i.e. characterized by cumulative growth of their ability to predict and control), whereas social sciences are not.[19] In consequence, economic actors facing the situation of choice cannot be sure what the consequences of alternative lines of action will be. Moreover, as mainstream economic theory admits, when we are faced with incomplete markets or imperfect information rational economic calculations may be impossible to perform.[20] In short, instrumental rationality, such as means-ends schema, can work successfully only in a highly predictable environment. But our social and economic world is not of this kind. Indeed, as many writers assert, unquantifiable uncertainty is one of the most enduring features of our contemporary economic life.[21] For that reason, as Dewey reminds us, "all action is invasion of a future, of the unknown".[22] Surely, economic action in an uncertain world cannot be adequately described as a maximization problem.

Another problem with mainstream economic theory is that in many situations we cannot distinguish strictly between means and ends. Suppose that a person wishes to attain a state of religious ecstasy.[23] It would be inappropriate to suggest that essentially the same outcome can be achieved more efficiently through the use of certain psychedelic drugs. In such situations, the use of means-ends schema tends to misinterpret the character of established social customs. As Michael Macpherson observed, "it may in some cases make more sense to conceive of cultural learning as supplying «ends to given means» rather than other way round" (1983, 108). The means-ends dualism assumes that all relevant means leading to projected outcome are known in advance. However, as Jens Beckert (2003) sought to demonstrate, creative economic action as personified by the figure of Schumpeterian entrepreneur, typically modifies existing patterns of conducting business. Hence, entrepreneurship cannot be adequately described as finding the most efficient means to preconceived ends. Following Dewey Jens Beckert argues that creative action reconstructs both traditional means and established ends and, for that reason, it cannot be adequately explained within the rational choice framework. Finally, for Dewey inquiry does not start with assigning means to ends but with the interpretation of some problematic situation, when we do not know what to do. Exploring the

---

[19] Cf. Hesse (1980).

[20] See Hahn (1980).

[21] On the role of uncertainty consult Davidson (2010).

[22] Dewey (1922, 38, 37).

[23] The example is drawn from Macpherson (1983, 108).

intricate complexities of the situation can suggest novel means as well as previously overlooked ends.[24]

Mainstream economics assumes that economic reality can be bifurcated into two entirely separate domains. On the one hand, there is a purely objective world of means, which can be studied with the aid of scientific methods. On the other hand, there is a purely subjective world of ends, which are introduced to the economy from outside by economic agents. It can be demonstrated that means-ends dichotomy that dominates modern economic thinking is a paradoxical consequence of the replacement of deistic naturalism of the seventeenth and eighteenth century with the image of science modelled after modern natural science. As Gunnar Myrdal observed, "the use of the categories of means and ends to order and arrange knowledge did not become important until political economy has outgrown the naïve philosophy of natural law" (1958, 209). For classical economists, such as Adam Smith, nature was inherently moral since it was a part of divine creation.[25] The natural order was supposed to work according to laws given by God. In Myrdal's terms essence of old naturalism was "direct identification of teleology with causality" (Viner 1927, 206; Myrdal 1958, 206). However, with the rise of modern physics classical deism fell out of grace and a different image of nature emerged. This time nature was depicted as a cold realm governed by impersonal mechanisms which were indifferent to human purposes. Thus, new naturalism modelled on the idea of physical nature has relegated causality and teleology, or means and ends, into two entirely separated realms. The result is extreme relativism, which characterizes mainstream economic thinking. All values are regarded as essentially subjective and all of them lay outside the scope of theoretical or positive economics.

Needless to say, the economic universe cannot be divided so neatly into means and ends. For instance, labour can be regarded both as a means (i.e. factor of production) and as an end because it is closely intertwined with the quality of life and dignity of human beings. Moreover, as we have already observed, economics errs in treating ends as given. The tendency to treat preferences as fixed is especially explicit in the economic theory of human behaviour put forward by Gary Becker (1976, 5). However, such a perspective is clearly at odds with many popular ideas about the nature of rationality and freedom. Especially for the normative purposes, rationality cannot be adequately conceptualized as a simple matter of obtaining the most efficient means for a fixed set of preferences. As John Dewey argued, "what men actually cherish under the name of freedom is varied and flexible growth, of change of disposition and character, that stems from intelligent choice" (1928, 110). Thus, the formulation of new and potentially more rational aims is far more important than finding optimal satisfaction to an existing set of preferences. Indeed, as Karl William Kapp observed, "only a sick or neurotic being, one who singles out one fixed end and attached an overruling importance to it, can be said to have a

[24] For a more elaborated account of the consequences of situational character of ends see Whitford (2002).

[25] Cf. Viner (1927).

given or fixed preference schedule" (2011, 79). Moreover, orthodox economic theory pays little attention to the conditions under which needs and preferences are formed. If the dominant set of preferences is created by giant corporations via means of advertisement and manipulative techniques of persuasion driven by big data technology, then it is not so great compliment to say that the market system is the most efficient way to satisfy those preferences. Clearly, what we would like to know is how autonomous people are in making decisions in the market and how rational their consumption choices are from both social and ecological point of view.

To sum up, the utilitarian account of rational choice, which underpins much of modern economic thinking, is inadequate both for the purpose of scientific explanation and moral guidance. Theory of rational action overlooks the role of institutions in shaping human conduct and rests on an overly simplified view of the relationship between means and ends. Moreover, it tends to miss what is arguably most important in moral deliberation and conduct, namely the possibility of rational assessment of diverging aims and conceptions of the good.

4.4 Embedded Markets and the Importance of Institutions

The last vestige of naturalism which I am going to discuss in this essay concerns the philosophical presuppositions of the language we often employ when talking about capitalism and markets. Naturalistic orientation in this area can be detected in the tendency to treat capitalism or markets as quasi-natural phenomena. Once set in motion, they tend to develop according to their own inner logic, which reflects their inner nature or essence. Interestingly enough, such views are entertained both by mainstream economists and their Marxist critics. John Dewey and Karl Polanyi were keen critics of the aforementioned tendency to treat capitalism and markets in a naturalistic manner. Writing to Clarence Ayres, Dewey expressed the hope that in the future, economics would offer an account of economic order that would "break down the idea that capitalism or any other socio-political order is fixed entity and hence exempt from otherwise universal sway of process".[26] For a similar reason, Karl Polanyi tended to avoid the word capitalism in his magnum opus *The Great Transformation*, preferring such notions as the market society.[27] At the same time, Polanyi was aware that the very idea of the market which we inherited from Adam Smith is wedded to naturalistic philosophy. One of the main contributions of Karl Polanyi to social thought is the recognition that markets are always socially constructed.[28] John Dewey on his own part offered a very similar critique of market naturalism. Writing about the need to reform liberal social thought Dewey argued

[26] Dewey's letter to Ayres quoted in Tillman (1998, 159).

[27] This point was noted early on by Allen Morris Sievers in first book-length study devoted to the analysis of Polanyi's thought. See Sievers (1949, 18).

[28] See Polanyi (2011).

against "the *laissez faire* doctrine held by degenerate school of liberals to express the very order of nature itself".[29] Both thinkers believed that overcoming market naturalism is a necessary prerequisite for realistic thinking about the place of markets in the modern society.

Such considerations may seem as one more curiosity pulled out from the rich treasury of bygone social and economic doctrines. Nevertheless, I would argue that our own way of talking about capitalism and markets is not far off from the assumptions criticized by Dewey and Polanyi. I will try to demonstrate that by discussing two recent books strongly influenced by Karl Polanyi's institutional theory: Fred Block's *Capitalism. The Future of Illusion* and Steven Vogel's *Marketcraft. How Governments Make Markets Work.*[30] Let's start with Vogel. His main thesis is that "real world markets are institutions: humanely devised constrains that shape human interactions" (2018, 1). To be sure, very few people would contest this claim. Even the most devoted followers of Hayek and Friedman are ready to admit that markets in order to function need a basic legal framework. However, they will also assert that when legal courts and property rights are in place markets can function in a self-regulatory fashion. However, this is precisely what Vogel denies. For him even the daily functioning of markets cannot be explained without accounting for cultural norms and practices, social networks, power relations, and public and private acts of governance. In other words, for Vogel markets are institutionalized and socially embedded almost to the bone.

One of the most interesting aspects of Steven Vogel's argument is close attention he pays to words we habitually use when discussing markets. In Vogel's view our economic language is still predominantly naturalistic. In our public discourse, we tend to treat markets as if they were natural phenomena. They key misleading phrase is the expression "free market", which was popularized by Milton Freedman. As Vogel explains this "elegant juxtaposition of the world *free* and *market* evokes many of the presumptions challenged in this book: that markets are natural; that markets arise spontaneously; that markets inherently constitute an area of freedom; and that government action necessarily constrains this freedom" (2018, 117). The misleading imaginary of "freeing" the markets is present in popular expressions such as leaving things to the markets or trusting the markets. It also manifests itself in the idea of government "interfering with" or "distorting" natural workings of the economy. Such abuses of language are by no means limited to popular discourse. Professional economics also uses rather problematic concepts such as perfect *competition* or *market failure*. The theory of perfect competition, as Vogel observed, "implies that imperfect markets are the puzzle to be observed and perfect markets are natural order" (2018, 122). In consequence, what economics tries to explain are deviations from market behaviour and not market behaviour itself (Vogel 2018, 122). In a somewhat similar vein market failure theory positions government as an outside force that corrects some minor dysfunctions of the economy. Yet such

[29] See Dewey (1935, 290).

[30] See Block (2018) and Vogel (2018).

conceptualization is misleading insofar as it suggests that failures are accidental to the workings of markets. However, flawless or perfect markets exist only in an imaginary realm of pure economic theory. More importantly, real markets would be inconceivable without rules, including legal rules sanctioned by the governments. Thus, the government cannot be portrayed as a force external to the market, which comes down to fix this or that occasional failure.

A similar point can be made also against the notion of capitalism. To be sure, we can define capitalism in purely institutional terms as a regime characterized by a dominant role of private property, prevalence of wage labour, and markets as main mechanisms for allocation of resources. However, it is important to realize that the notion of capitalism is very often associated with the image of a fixed system driven by its own inner laws. As Fred Block argues, this invokes the image of capitalism as a natural organism with his own DNA (2018, 28). From this perspective attempts to reform capitalism can be depicted as contradicted by the inner logic of the system or inconsistent with its DNA. But such arguments are both misleading and dangerous. It is misleading to think of capitalism as a coherent system. As Fred Block argues, "market economies depend on complex combination of conflicting institutions and motivations, they are contradictory and unstable, and they periodically require major structural reorganizations" (Block 2018, 15). More importantly, the view that capitalism is an autonomous system which follows its own inner logic puts unnecessary constraints on democratic political action. The image of capitalism as driven by its own laws suggests that democratic governments have no choice but to conform to its systemic demands.

To sum up, I have argued that contemporary public discourse about capitalism and markets is contaminated by unexamined naturalistic assumptions. Naturalistic vocabulary is omnipresent both in public debate and in the realm of academic social theory. John Dewey's pragmatism and Karl Polanyi's institutional theory can be seen as parallel attempts to draw our attention to the fact that markets are not governed by natural laws of economics but rather by man-made legal and political rules. Hence, we can always attempt to rewrite those rules in the face of new social, economic and environmental problems.[31]

4.5 Concluding Remarks

In this paper, I have argued for a particular set of alliances between different philosophical approaches to social inquiry and rival research programs in contemporary social sciences. I have argued that mainstream economics is closely associated with positivist currents in philosophy of science. I also noted a series of elective affinities between various forms of institutionalism and such philosophical positions as

[31] For a very influential attempt to apply institutional insights about the importance of rules to the problems of contemporary capitalism see Stiglitz et al. (2016).

pragmatism, constructivism, and hermeneutics. It may well be true that the growing discrepancy between mainstream economics and institutionally oriented social sciences is one of the most profound divisions not only in academia but also in contemporary Western culture writ large. From the historical perspective, it can be seen as a modern version of the *Methodenstreit* in which different philosophical orientations intersect with methodology and politics. In case of such deep-seated cultural conflicts, it is unlikely that the debate can be resolved in any predictable future. At any rate, this paper does not claim to offer knockdown arguments against mainstream economics. Its more modest ambition is to persuade the followers of an institutional approach that their arguments can be supplemented with valuable insights derived from Dewey's pragmatism, social constructivism and hermeneutics. By incorporating such a philosophical perspective, proponents of institutionalism can not only find new arguments against their neoclassical rivals but also better understand philosophical presuppositions of their own theories.[32] Institutionalism is most persuasive when we recognize that the economy is constituted by socially constructed rules and that those rules are being permanently reconfigured in response to various difficulties and contradictions, which Dewey once called problematic situations. The aim of economic inquiry is to aid democratic societies in this endless process of institutional reconstruction.

References

Arnsperger, Ch., and V. Yannis. 2008. Neoclassical economics. Three identifying features. In *Pluralist Economics*, ed. Edward Fullbrook, 13–25. London/New York: Zed Books.

Becker, G. 1976. *The economic approach to human behaviour*. Chicago/London: University of Chicago Press.

Beckert, J. 2003. Economic sociology and Embeddedness. How shall we conceptualize economic action? *Journal of Economic Issues* 37 (3): 768–787.

Bernstein, R.J. 1999. *Praxis and action. Contemporary philosophies of human activity*. Philadelphia: University of Pennsylvania Press.

Blakely, J. 2016. *Alasdair MacIntyre, Charles Taylor and the demise of naturalism. Reunifying political theory and social science*. Notre Dame: University of Notre Dame Press.

Block, F. 2018. *Capitalism. The future of an illusion*. Oakland: University of California Press.

Bohman, J. 2002. How to make a social science practical: Pragmatism, critical social science and multiperspectival theory. *Millennium: Journal of International Studies* 31 (2): 499–524.

Dale, G. 2016. *Karl Polanyi. A life on the left*. New York: Columbia University Press.

Davidson, P. 2010. Risk and uncertainty. In *The economic crisis and the state of economics*, ed. Robert Skidelsky and Christian W. Wigström, 13–29. New York: Palgrave Macmillan.

Dequech, D. 2007. Neoclassical, mainstream, orthodox and heterodox economics. *Journal of Post-Keynesian Economics* 30 (2): 279–302.

[32] It should be noted that the argument linking institutionalism with pragmatism, constructivism and hermeneutics is not the only possible one. For instance, Tony Lawson has offered interpretation of institutional economics, which attempts to tie it to a realist ontology. See Lawson (2005).

Dewey, J. 1915/1967–1990. The logic of judgments of practice. In *The collected works of John Dewey, 1882–1953*, ed. Jo Ann Boydston 14–97. Carbondale/Edwardsville: Southern Illinois University Press.
———. 1922/1967–1990. Human nature and conduct. In *The collected works of John Dewey, 1882–1953* 14, ed. Jo Ann Boydston. Carbondale/Edwardsville: Southern Illinois University Press.
———. 1927/1967–1990. The public and its problems. In *The collected works of John Dewey, 1882–1953*, ed. Jo Ann Boydston. Carbondale/Edwardsville: Southern Illinois University Press.
———. 1928/1967–1990. Philosophies of freedom. In *The collected works of John Dewey, 1882–1953*, ed. Jo Ann Boydston, 92–114. Carbondale/Edwardsville: Southern Illinois University Press.
———. 1929/1967–1990. The quest for certainty. A study of the relation of knowledge and action. In *The collected works of John Dewey, 1882–1953*, ed. Jo Ann Boydston. Carbondale/Edwardsville: Southern Illinois University Press.
———. 1935/1981–1990. The future of liberalism. In *The collected works of John Dewey, 1882–195*, ed. JoAnn Boydston, Carbondale: Southern Illinois University.
———. 1938/1967–1990. Logic. The theory of inquiry. In *The collected works of John Dewey, 1882–1953*, ed. Jo Ann Boydston. Carbondale/Edwardsville: Southern Illinois University Press.
———. 1947/1967–1990. Liberating the social scientist. In *The collected works of John Dewey, 1882–1953*, ed. Jo Ann Boydston, 224–238. Carbondale/Edwardsville: Southern Illinois University Press.
Hahn, F. 1980. General equilibrium theory. *The Public Interest* 36: 123–138.
Hesse, M. 1980. *Revolutions and reconstructions in the philosophy of science*. Bloomington/London: Indiana University Press.
Hodgson, G. 2002. Institutional blindness in modern economics. In *Advancing socio-economics. An Institutionalist perspective*, ed. J. Rogers Hollingsworth, Karl H. Müller, and Ellen Jane Hollingsworth, 147–170. Lanham/Oxford: Rowman and Littlefield.
Joas, H. 1996. *The creativity of action*. Chicago: The University of Chicago Press.
Kapp, K.W. 2011. In *The foundations of institutional economics*, ed. Sebastian Berger and Rolf Steppacher. London/New York: Routledge.
Lawson, T. 2005. The nature of institutional economics. *Evolutionary and Institutional Economics Review* 2 (1): 7–20.
———. 2006. The nature of heterodox economics. *Cambridge Journal of Economics* 30: 483–505.
Lewis, C.I. 1939. Meaning and action. *The Journal of Philosophy* 36 (21): 572–576.
McPherson, M. 1983. Want formation, morality and some interpretive aspects of economic inquiry. In *Social Science as Moral Inquiry*, ed. Norma Hann, Robert Bellah, Paul Rabinow, and William Sulivan, 96–124. New York: Columbia University Press.
Mills, C.W. 2000. *The Sociological Imagination.*. Afterword by Tod Gitlin. Oxford/New York: Oxford University Press.
Mirowski, P. 1986. Mathematical formalism and economic explanation. In *The reconstruction of economic theory*, ed. Philip Mirowski, 179–240. Boston: Kluwer-Nijhoff Publishing.
———. 1989. *More heat than light. Economics as social physics: Physics as nature's economics*. Cambridge: Cambridge University Press.
Mitchell, W.C. 1950a. The quantitative analysis in economic theory. In *The backward art of spending money and other essays*, ed. Wesley Mitchell, 72–82. New York: Augustus M. Kelly Inc.
———. 1950b. The prospects of economics. In *The backward art of spending money and other essays*, ed. Wesley Mitchell, 342–386. New York: Augustus M. Kelly INC.
Myrdal, G. 1958. Means and ends in political economy. In *Value in social theory. A selection of essays on methodology*, ed. Paul Streeten, 206–231. London: Routledge and Kegan Paul.
Polanyi, K. 1977. *Livelihood of man*. New York/San Francisco/London: Academic.
———. 2011. *The great transformation. The political and economic origins of our time*. Boston: Beacon Press.

Putnam, H. 2002. *The collapse of the fact-value dichotomy and other essays*. Cambridge, MA/London: Harvard University Press.

Robbins, L. 1932. *An essay on the nature and significance of economic science*. London: Macmillan & Co.

Róna, P. 2017. Why economics is a moral science? In *Economics as moral science*, ed. Peter Rona and László Zsolnai, 3–9. Cham: Springer.

Searle, J. 2005. What is an institution. *Journal of Institutional Economics* 1: 1): 1–1):22.

Sievers, A.M. 1949. *Has the market capitalism collapsed?* New York: Columbia University Press.

Stiglitz, J., Abernathy, N., Hersh A., Holmberg S., Konczal M. 2016. Rewriting the Rules of the American Economy. *An Agenda For Growth and Shared Prosperity*. New York/London: W.W. Norton & Company.

Taylor, Ch. 1971. Interpretation and science of man. *Review of Metaphysics* 25 (1): 3–51.

Tillman, R. 1998. John Dewey as a user and critic of Thorstein Veblen ideas. *Journal of the History of Economic Thought* 20 (2): 145–160.

Viner, J. 1927. Adam smith and laissez faire. *Journal of Political Economy* 35 (2): 198–232.

Vogel, S.K. 2018. *Marketcraft. How governments make markets work*. Oxford: Oxford University Press.

Westbrook, R.B. 1991. *John Dewey and American democracy*. Ithaca, NY: Cornell University Press.

Whitford, J. 2002. Pragmatism and the untenable dualism of means and ends: Why rational choice theory does not deserve paradigmatic privilege. *Theory and Society* 31 (3): 325–363.

Chapter 5
Moral Economics – A Theoretical Basis for Building the Next Economic System

Zsófia Hajnal

Abstract This paper explores the concept of moral economics in the author's own, complex interpretation. Moral economics is a new approach to the structure and nature of the economic system, suggesting changes at several critical points. The paper is an interdisciplinary work between philosophy and economics, shifting from theory to practice, past to present, and balancing between the normative and the positive. The purpose is to change the economic understanding and the perceived economic logic through incorporating the moral factor into comprehensive models.

5.1 Introduction

This paper aims to give a sketch – on this scale and in this scope, it can only be called a sketch – of moral economics and of the moral economy. Moral economics is not a well-established economic philosophy as of yet. In more recent interpretations, it is a new school of economic thought, still searching for and paving its own ways and opportunities.

The moral economy interpretation of this paper aims to provide smoothening and completing elements to the existing system. However, moral economics also aims to redefine economics as we know it today, by adding the moral factor to its implications. The new concept emphasizes the finite nature of humankind's resources, the interests of the community beside the interests of the individual, takes into account the needs of future generations, and calculates with giving other species space and opportunity to live (Tóth 2016, 38). In the current interpretation, the points to be highlighted most should be that – as a philosophy of economics – it is organized for sharing, and that it includes rational selflessness in its mechanisms.

Z. Hajnal (✉)
International Relations, Corvinus University of Budapest, Budapest, Hungary

International Economics and Business, Budapest Business School – University of Applied Sciences, Budapest, Hungary

P. Róna et al. (eds.), *Words, Objects and Events in Economics*, Virtues and Economics 6, https://doi.org/10.1007/978-3-030-52673-3_5

Not making this connection between morality and economics could cause harm. *"In fact, separating morality from economics, which seems like a disciplinary "purification," is actually the regression in the development of economics."* (Wang 2015, 84).

5.1.1 Outline and Method

In terms of research questions, there are multiple goals. The questions can be grouped into four main categories, whereupon each main chapter responds respectively.

First, the questions regarding the taxonomy of moral economics: Where are its roots? How did the term shape and change its meaning during the past centuries? How can it be applied at present? Where can moral economics be located on the shaded map of the economic systems?

Second: Is the moral structure of society and the resulting mechanism corresponding to the Smithian view and its interpretation? Is the bottom-up system of capitalist mechanisms sustainable for modern economies? Is it reasonable to build an economic system on individual selfishness?

Third, the questions aimed at the apparatus of moral economics: What are the mechanisms and structures that would make it work? What does equilibrium mean in moral economics, or more accurately, where is the point of equilibrium to be reached in this proposed new context?

And fourth but not least, the question connecting the previous findings to the present: What are the current tools of migrating into the moral economic system?

Throughout the paper there is a shift from theory to practice orientation, as well as from a future-view to a focus on the present. Further, it is neither clearly positive, nor purely normative. The attempt to establish a unified and complete system of moral economic thinking tilts it slightly into the positivist direction, but its context and place on the economic spectrum lends it a normative nature. The normative nature should not be perceived as a weakness, as contemporary economists, such as Tomáš Sedláček (2012, 21), argue for the normativity of economics itself.

5.1.2 Baseline Assumptions

The need for cooperation is rather a feature of the community, and extrinsic to individuals, stemming from the lifestyles they created for themselves. The need to help others, to respect others and to lift them up, on the other hand, is intrinsic to individual human mechanisms. This need can be called rational selflessness. One aim of the paper is to demonstrate how economics works better, i.e. how its models form a more comprehensive system, if the moral factor is incorporated, both on the individual level, and in terms of mechanisms.

The paper describes economic orderliness as follows: A community has economic orderliness, if the needs of the individuals are fully satisfied, without exceeding the individual need-targets, i.e. without wasting resources, and if satisfying needs is secured for the future by interacting sustainably with the environment.

The novelty of this definition lies in the emphasis on needs and precision. This type of precision implies effective distribution. The first part foreshadows the requirement for quantifying human needs. The second part of the definition is not novel, yet it has to be stated as a prerequisite for economic orderliness.

Moral economics holds a non-hedonist view about the overall good. It is not the pleasure in itself that needs to be maximized, but, rather, order. This order has three components that are rules going against present-day tendencies:

First, on an individual level, the goodwill for the whole has to overcome the ego, i.e. people have to become conscious of the causes and consequences of rational selflessness. Individuals should have enough information, an overview and an understanding of the system, so that they recognize the benefits of sharing assets and joys with fellow humans. Secondly, pleasure of any type is more valuable if it is given to those more in need of it. As Tóth (2016, 200) writes: *"Material growth of the strong at all costs should not be replaced by de-growth or zero growth, but by the growth of the weak, as their utility – and thus the objective utility of those in more advanced situations – is greatly increased by their material growth"*[1] Thirdly, and finally, during all the production and distribution, the environment has to stay in a sustainable state.

To summarize the above paragraph: Human economic consciousness has three levels in moral economics, and only a system in which all the three rules are kept, and selflessness is expanded to the whole, can be called a moral economy.

5.1.3 The Approach

As moral economics in the broader context is a relatively new theory, it has to be stated that the current interpretation is that of the author of this paper. It builds partially upon relevant literature, but goes beyond that in many aspects, beginning with the basics of the approach.

The paper has five underlying axioms not all of which have scientific proof at present. In certain cases, we are talking about concepts having been debated already for thousands of years, and it is not within the scope of this paper to give proof or answers to the related philosophical questions. However, the axioms need to be enlisted, as they are necessary for sketching the moral economy.

Firstly, morality's objectivity is assumed as a premise. Although this assumed objectivity does not give direct aid in certain moral dilemmas or lawsuits, it

[1] Own translation. Original in Hungarian: *"az erősek mindenáron való anyagi növekedése helyére nem a csökkenés vagy zéró növekedés kerül, hanem a gyengék növekedése, hiszen az ő hasznosságukat – és ezzel a fejlettek objektív hasznosságát – nagymértékben emeli az anyagi növekedés"*.

strengthens the intuitive stance that moral advancement is meaningful, and that progress has a certain direction. With regard to morality's definition, the paper borrows from Wang Xiaoxi (2015, 56), a contemporary Chinese economist, the following sentence: *"Scientifically, ethics has the function of encouraging people to constantly improve themselves, and at the same time, continually cherish and improve the interpersonal relationships within co-existence, so as to build a better living environment based on the concept of rational existence, pushing forward sustainable social development."*

Secondly, humankind is seen now as an entity that constitutes one big community. Not only among historians is this evident, but also for the CEO of the largest social media corporation: *"History is the story of how we've learned to come together in ever greater numbers – from tribes to cities to nations. (…) Today we are close to taking our next step."* (Zuckerberg 2017). Beside the social aspect, this statement is also valid for economics. However, a healthy balance is carefully intended to be kept between the individuals' and the community's emphasis, throughout the entire paper.

Thirdly, it is assumed that humankind has a goal: to increase human well-being. It is being achieved by progress that is linear to some extent.

The fourth premise is that of rational selflessness. It is an intrinsic human need of certain altruism. This need is definitely quite high up on our needs pyramids, but it does exist and influence humans. To confirm the existing relationship between the needs pyramids and morality, Wang (2015, 68) should be quoted again: *"man has his own different kinds of unique pursuits, among which, those at higher levels are all related to mortality".*

Fifth but not least, this paper takes the stance that individual human needs are not infinite, nor are they insatiable. Growing expectations may be experienced when needs on higher and higher levels get satisfied, but these expectations are just recognitions of needs which already existed and they do have a rational end, which would be more visible in an orderly and just world, tailored to people's needs.

These axioms may be a matter of faith, but some voices already echo that so are the axioms of mainstream economics (Tóth 2016, 198; 284). The special reason for that these axioms had to be stated is their intuitive nature. Although – as will be seen in the forthcoming chapters – the moral economic models, principles and suggestions are intended to be built up rationally, the system stands on intuitive axioms, mostly.

5.2 Conceptual History and Ideological Context

The following two chapters delve into the conceptual history of moral economics and its ideological context. In the former, the paper examines roots and historical appearances of the concept, in the latter, whether it is suitable for being the next paradigm. This is a crucial part, because it also helps to understand what the current concept of moral economics distances itself from, and what it is debating, or at least being critical of.

5.2.1 A Brief History of the Moral Economic Concept

The concept of the moral economy and of moral economics used in this paper is not to be confused with the concept which is examined in the context of eighteenth century peasant economies, and for which we find the first results when searching by the words "moral economy" on the Internet.[2] The term moral economy has been brought to the historical agenda by the British historian E. P. Thompson in a 1971 article *(The Moral Economy of the English Crowd in the Eighteenth Century)*, but the way he used it, it remained *"bound to a specific epoch and a particular historical context"* (Götz 2015, 147). According to the Swedish historian Norbert Götz, *"the concept* [the moral economy] *has the potential of improving the understanding of modern civil society."* Götz also expressed his concern of Thompson's concept of the moral economy having in fact no moral implications, while Thompson himself was *"concerned about the conceptual preservation of the historical context that he had assigned to the term and about a possible loss of specificity upon its free adaptation by others."* (2015, 153). This indicates a rather significant tension between near-past and present interpretations, but tensions exist even amongst the present-day versions of the moral economy.

For example, this paper separates itself from those moral economic concepts which have their emphasis on religion or emotions, without directly linking them to or embedding them into economics. The relevant Hungarian literature on *emotion-economics* and *humane economics*[3] can be named, written about by the economists Balázs Hámori and Gergely Tóth respectively. According to the author of this paper, moral economics should be a broader term, as the aforementioned approaches *"do not address economic issues in the way they are commonly understood"* (Götz 2015, 147).

The term *moral economy* has already had several other appearances throughout history. The two words are stemming from Latin and ancient Greek respectively. As per Götz, the word *moral* did not need to be added to *economy* until the middle of the eighteenth century, as the word *economy* contained morality self-evidently. The very first emergence of the compound is believed to have taken place in 1729, *"in a sermon preached before the University of Cambridge"*. The time the term appeared second was in the *Athenian letters*, published in 1792 (circulated privately about fifty years earlier already): It *"was used in a letter purportedly written by the Persian King Smerdis corresponding with his agent in Athens at the time of the Peloponnesian War"*, referring to deity, similarly to the first occurrence. The third (Anglo-Saxon)

[2] This search was carried out through Google on the 2nd of September 2017.

[3] Translated by the author from the Hungarian terms *"érzelemgazdaságtan"* and *"humánökonómia"*.

appearance came about in 1762, in a poem of Amyas Bushe, celebrating *"the harmony of the human will that was equally determined by reason and passion"*. Meanwhile, Rousseau discussed moral economy in his *Encyclopédie* article, with a similar connotation of justice and balance. An explicit meaning of the term *moral economy* was applied by other authors, e.g. Fortunato Bartolomeo de Félice (1769), Louis-Claude de Saint-Martin (1774), Johann Friedrich Wilhelm Jerusalem (1774) and Johann Friedrich von Ungern-Sternberg (1785), who *"contrasted the physical economy of all creatures with the moral economy particular to the human condition"*. So did the *Encyclopædia Britannica*, in its third (1797) edition. *"By the end of the eighteenth century the term 'moral economy' became part of the title of a French dictionary distinguishing political, civil, and moral economy (Beauvray, 1770)* (Götz 2015, 149)."

Later on in the eighteenth century, *"the French administration in the Rhineland (...) expected seats of higher education to establish 'un Professeur d'économie morale'"*, which was rather an evolutionary exploitation of the concept. Staying at the topic of the French Revolution, *"Michio Shibata frequently used moral economy when discussing food riots in the ancient régime and the Parisian sans-culottes in the revolution."* Also, the economist Jean Herrenschwand *"deduced ideas about the evolution of increasingly advanced economic systems" "in his principal work, De l'économie politique et morale de l'espèce humaine"* (1796) (Götz 2015, 150). It is worth mentioning Adam Smith at this point, who has not used the term moral economics himself, but whose legacy is occasionally spoken of by some of his disciples and his modern exegetes as his *moral economy* (Götz 2015, 153–154).

There were several religious interpretations of the moral economy in the eighteenth and nineteenth centuries, and it was the radical reformers who employed the concept to the largest extent (Götz 2015, 150–151). From the early nineteenth century on, the term moral economy *"could be associated with both capitalism and socialism* (Götz 2015, 151). The next chapter will expound on the relations of these ideologies and economic paradigms with moral economics in more detail.

After the publication of E. P. Thompson's 1971 article *"references to moral economy have proliferated [...] as a slogan of critics of the market system"* and many more extended notions have been established, in the non-profit sector as well. Whether the emphasis is on the word "moral" or on "economy" varies (Götz 2015, 155–57).

Science fiction has produced a variety of moral economic worlds, e.g. through the pen of the author Thomas Dick (Götz 2015, 151). The space civilization called *Mül* from the French science fiction comic series *Valérian and Laureline* (1967–2010) also has certain features of moral economics. Later on in the paper it will be shown that although humankind does not have *Mül converters* (creatures being able to replicate anything they eat), an orderly economy in harmony with nature (such as that of *Mül*) is feasible for humans too.

None of the enumerated approaches and interpretations correspond to the moral economic concept of the current paper. Attempts to broaden the meaning of the moral economy are only of recent origin. There is no group in the academic sphere yet, who would call themselves moral economists or representatives of the moral economy. The term moral economics has not officially been used in the academia so far, as the mechanisms and a comprehensive economic description have not been elaborated on yet.

The moral economic concept in this paper is not a reference to certain economic relations, or any particular dimension or sector of the economy. It is a synonym for the ideal structure of the economy, constructible by humankind, so it is encompassing a larger system than the concept of the modern market economy.

5.2.2 Ideological Context and Positioning

Moral economics is a new economic paradigm applicant, a position for which there are not many applicants. Even the contemporary Polish economist Grzegorz Kolodko's *Chinism* (Kolodko 2018, 23) is claimed to be a transition period only, no matter how long it will last.

For many decades, the economic profession has excluded (or at least not seriously considered) opportunities of third (or fourth, etc.) variants to be economic paradigms. The reason for this may lie in the nature of the alternatives, which have actually never stepped out of the socialism-capitalism dichotomy. Ideological debate has mostly been about choices between the two (with significant systemic shortcomings) and trade-offs of their different variations. Moral economics has more dimensions: explanations, principles and solutions (which will be presented later on) that are solving or dissolving its predecessors' problems, at least in theory.

At present times, China is a self-proclaimed socialist system (Wang 2015, 57) yet it has mixed features of, and even distinguishing elements from the two major economic system-types. Socialism and capitalism, two mammoth-ideologies of economics and politics (and even culture), have enormous literature,[4] and for most of the twentieth century it was these two which *"functioned and confronted each other in practice"* (Kolodko 2018, 2).

Regarding the economic ideologies, recent generations have not inherited a consensus over the different meanings attributed to them. In fact, there is still a *"confusion in definitions and the lack of methodological discipline"* (Kolodko 2018, 2).The reason for this is to be sought (partially or mostly) in the differing variations of these economic regimes, not only in space, but also in time (Kolodko 2018, 3), which already raises two questions for moral economics. Namely that if it is to be the next economic system, should it be universal, or universal only to a certain extent, and

[4]The best known contemporary Hungarian economist, János Kornai, has a life's work on these topics.

whether the universality of an economic system would be an obstacle for its implementation in terms of the respect for local culture (the word local could encompass whole continents here, as it may stand for certain regions). In fact, the paper does not go as far as implementation. The moral economy may at the end of the day be a better or a worse system than theoretical moral economics, scientific moral economics, or – one dare say – textbook moral economics have ever prescribed.

Socialism in the twentieth century in general was too ambitious compared to the historical settings, its technological means and the available knowledge. For moral economics, all three factors have changed. The distance between the present situation and the declared goal is estimated to be measurable in units of decades or even only years.

Kolodko (2018, 2) describes capitalism very shortly compared to its scope as *"a socio-economic system based on private capital aspiring to maximize its profits"*, but even this concise definition is in a major contradiction with moral economics, which is based on needs (including the need to help others) and aspires to maximize well-being rationally.

In his conference paper about Chinism, Kolodko (2018, 3–6) enlists a number of economic system variations from the socialism-capitalism range (or rather plane). Just to name a few examples: classical capitalism, ideal communism, socialism with Chinese characteristics, state capitalism, the ideal social democracy, the social market economy of Scandinavian countries, post-communism, post-Soviet state capitalism, emerging markets with a socially-oriented capitalist economy and contemporary capitalism. The variables of his categorization were borrowed from the late Italian economist Mario Nuti, and are as follows: dominant public property and enterprise, equality and large public consumption, economic democracy and participation, social control of the main economic variables (employment, income, accumulation, growth, inflation, internal balance, external balance). If we ignore some dimensions of moral economics, and try to force it onto this scheme, we may say that this perpendicular projection stands closer to the existence of all these factors than to their nonexistence. Based on the strict evaluation (Nuti's method of categorization) of these factors, the moral economy's projection is 37.5% capitalism and 62.5% communism, but these numbers are illustrative, and are not suitable for drawing conclusions.

Moral economics is no different from other ideologies in that it aims to create the perfect economic frame for human coexistence and cooperation. It is new in that however – as this paper claims – it is the first category on the extended (multi-dimensioned) taxonomic map of economic systems that has found the models constituting the perfect frame, and although the statement is not empirically proven on the larger scale yet, most of the upcoming chapters are contributing to its verification, while leaving an open door for falsifiability, down to the system's axioms and *raison d'etre*.

5.3 Adam Smith Revisited[5]

This chapter aims to provide a fresh approach to Adam Smith's moral and economic ideas, relying to a certain extent on the original works, *The Theory of Moral Sentiments* (1759) and *An Inquiry into the Nature and Causes of the Wealth of Nations* (1776), but also taking new secondary sources into account. Some of the broadly accepted, interpreted (and misinterpreted) ideas are challenged, in the context of the international political economy discipline.

Adam Smith (1723–1790) was definitely at the right time (at the dawn of industrialization, and the Scottish Enlightenment) in the right place (Scotland, Britain) with his ideas, but presently, these ideas need to be revisited. His ideas have been very influential, but are still controversial, often just cherry-picked by philosophers and politicians from both sides of the political spectrum (BBC World Service 2017).

The research questions for this chapter start out from a chapter of Varoufakis' book *Foundations of Economics: A Beginner's Companion* (1998, 16–21). In both of his major works, Adam Smith attempted to justify human systems (that of morality and that of the economy) through inherent mechanisms of the individuals and their interactions. The two models are actually similar in their framework, but not necessarily flawless.

5.3.1 The Moral Structure of Society

In early political economy it is common to see references to morality (Sayer 2000, 84). In his work *The Theory of Moral Sentiments* – which is in the views of many rather of psychological character than of philosophical – Adam Smith built a moral system based on individual judgements and empathy (incorporating the notion of the *"impartial spectator"* in his model), *"rather than beginning from a philosophical vantage point above those judgements"* (Stanford Encyclopedia of Philosophy 2013). He adopted a bottom-up approach (BBC World Service 2017), a method which he exercised in *The Wealth of Nations* as well. Still, Smith was a moral universalist, who followed the Stoic tradition, thinking *"that our moral feelings extend, if to a lesser degree, to all rational and sensible beings"*, and who aspired to provide *"a structure for morality that reaches out across national and cultural borders"* (Stanford Encyclopedia of Philosophy 2013).

The moral picture and stance regarding society in *The Wealth of Nations* seems to be in contrast or disconnected from that in *The Theory of Moral Sentiments*. In Varoufakis' interpretation of the former, Smith's pragmatist attitude suggested that

---

[5] This chapter is largely based on an essay from the author of this paper, with the title *Adam Smith Revisited – An Inquiry into the Present-day Relevance and Applicability of his Major Works*, handed-in for the subject *International Political Economy*, 2017/18, I. semester at the Budapest Business School – University of Applied Sciences.

it is not in the (direct) interest of the merchant and capitalist class to contribute to the good of society. They are guided by greed, and if it was not for the automatic coordination, they would not take the actions that enhance public good (Varoufakis 1998, 18). Smith understood self-interest in *The Wealth of Nations* differently than in *The Theory of Moral Sentiments*. In the latter, he appeared to understand that selfishness is a misconception of self-interest, whereas in the former, he abandoned the moral structure. In *The Theory of Moral Sentiments* Smith is aware of the superiority of morality, in which *"a virtuous agent sees things that others do not"* and where morality *"is a way of co-operating with the Deity"* (Stanford Encyclopedia of Philosophy 2013).

In a world that is growing closer together, but is also quickly growing apart through its unprecedented interconnectedness, new levels of moral consensus rise, whilst previously hidden differences in people's views come to light. Still, at present day, humankind has a broader knowledge about morality, as well as a more elaborate picture of the moral structure of society and of entrepreneurial morality, than back in Smith's times. The general worker (or employee), who is also a consumer, has a better opportunity to see how he contributes to the flows of the economy. This type of inclusiveness and the feeling of being part of something bigger make it easier to be compliant with corporate guidelines, or the economy in general. In addition, boundaries between today's "capitalists" and the employees are fading on multiple frontiers, for example through promotion and shareholdership at the workplace, or through broadly available opportunities of entrepreneurship. Furthermore, through the – by now almost mandatory – CSR approach, as well as their mission statements, firms of different sizes have started to cultivate their moral capital. In moral economics, this is considered to be a step forward theoretically, no matter whether it happens impeccably in practice or not. Modern entrepreneurs openly express how their businesses operate for the good of society and many of them engage in philanthropic activities.

5.3.2 The Durability of a Bottom-Up Economic System

Economic systems have boundaries from below and above, thus the Smithian automatic coordination has its limits (Varoufakis 1998, 17). This holds true even in the current market economy. Regarding the lower boundary: the economic mechanism depicted by Smith is not effective enough to reach those on the bottom of society. As for the upper boundary: environmental harm would be escalated through pure capitalism. The economy as a growth machine is externally not sustainable.

It is not uncommon in economic philosophy, that the answers generate further questions. The question arising from the previous paragraph is whether – with the required measures taken – Smithian capitalism is suitable for the economic realm within the mentioned boundaries. This is not a hypothetical question, as developed countries are – from this viewpoint in slightly different ways – trying to tackle the global problems beyond the lower and upper boundaries, i.e. poverty and climate change. Between those boundaries (and in reality extending over them) their markets function in capitalist structures.

Economic durability lies in a state of equilibrium. On an ever-perfect market, Smithian capitalism is leading to the equilibrium of supply and demand. But the problem lies not in market imperfection. It is the wrong choice of factors for equilibrium. A bottom-up economic system, where supply and demand meet each other, may have the value of freedom of choice, but is not completely secure on the macro-scale. If, however, the choice of factors for equilibrium falls upon the needs of the population, and upon their potential output, a more complete and perfect economic system can be created. This is a top-down approach, and as history has proven (through several failures), humankind has technically not been ready so far for the shift. This paper aims to make a point for humankind to rely less on the Smithian *unintended consequences* in economics. Under present-day circumstances it is more reasonable to cultivate states' moral functions than denying these functions and keeping the role of the state minimal. From today's perspective, Smith's political views may seem distorted. *"He believed that states could and should re-distribute wealth to some degree, and defend the poor and disadvantaged against those who wield power over them in the private sector"*, yet he was sceptical towards progressivism and *"suspicious of large-scale plans for the reform of society"*. *"He was one of the earliest and most fervent champions of the rights and virtues of the poor"*, but his *"writings are permeated by a lack of respect for the sorts of people who go into politics"* (Stanford Encyclopedia of Philosophy 2013). Derived from the quoted contradictions, it is not possible to create a harmonious social and economic system solely based on Smith's legacy.

5.3.3 Wrong at Its Roots – What the Broader System Is Actually Built Upon

Adam Smith discovered several features of society's economic design (the power of labour division, supply and demand dynamics, and the *invisible hand*), but in the explanation, he laid the wrong foundation stone. This paper argues that the economic mechanism of the whole system (i.e. society, national and global) is not built upon selfishness, but on needs. Selfishness is a negative trait, whereas needs are natural phenomena, and thus neutral.

The world is very complex from the needs aspect. At the moment, in an extended need-sense, most of humankind is still just trying to survive. However, many in the developed world have realised that humankind would get faster over this problem if they cared about those most in need. This recognition can be traced back to sufficient knowledge and information, which most people in Adam Smith's time, and even Smith himself may not have had.

A system built primarily on selfishness will fail on the long term. Proof for this are models of game theory (Adami and Hintze 2013) as well as scientific alternatives to the *homo oeconomicus* (Alger and Weibull 2013). Selfishness does not pay off on the long run, and the *homo moralis* has better chances of success in evolution, than the homo oeconomicus.

Varoufakis (1998, 21) writes in his interpretation of Smith that *"as capital accumulation and economic growth gather pace, the inequalities between classes will shrink; people from different backgrounds will start moving closer together on the ever rising escalator."* In the twenty-first century, other tendencies are to be observed. The scissors are not closing – they are opening: *"How can we decide whether there is a net gain to society? In his invisible hand statement, Smith refers to the annual revenue of society or the national income, as we would say today. This indicates that society gains if the winners in the process of structural change gain more than the losers lose. But this is not an entirely convincing argument. Suppose that those who gain are already well off while the losers live in poverty. Would we not in this case hesitate to say that the invisible hand of the market works in the interests of society? And if so, what principles should guide our aggregation of individual interests into a measure of the interest of society as a whole?"* (Sandmo 2014, 7).

By now, economics has well-grounded evidence and explanation for the opening scissors in social wealth. As Thomas Piketty (2014, 571) asserts in the Conclusion of *Capital in the Twenty-First Century*: *"The principal destabilizing force has to do with the fact that the private rate of return on capital (…) can be significantly higher for long periods of time than the rate of growth of income and output (…). The inequality implies that wealth accumulated in the past grows more rapidly than output and wages. (…) The entrepreneur inevitably tends to become a rentier, more and more dominant over those who own nothing but their labor. Once constituted, capital reproduces itself faster than output increases"*.

Be it said in Adam Smith's defence: His ideas were consistent in the context of his own times, the times of small manufacturers, and competition seemingly becoming ever more perfect.

5.4 Mechanisms and Structures

At this point the paper turns to the actual micro – and macro-level models of moral economics. In the first three chapters, there is an arc to be observed, along which moral economics rebuilds economics itself. It goes back to the very basics, and has a different starting point. It starts from the selflessness of the individual, and builds the system up to the dynamic equation of aggregate needs and output, through introducing the sharing multiplier. The consequent three chapters expound on these models, and fine-tune the system created.

5.4.1 *Homo moralis* and Rational Selflessness

In the context of this paper, the concept of homo oeconomicus is a marginal one. This view is borrowed from Ingela Alger and Jörgen W. Weibull (contemporary economists), more specifically from their paper *"Homo moralis – preference evolution under incomplete information and assortative matching"*. Therein they write that *"the more general notion of homo moralis (…) should replace the more special notion of homo oeconomicus (with zero degree of morality) as a benchmark for human motivation"* (Alger and Weibull 2013, 29). The two researchers are not alone in opposing homo oeconomicus to be the general economic motivational notion. However, it is their understanding of the individual as an economic being that stands closest to this paper's stance: *"We call individuals with such preferences homo moralis and the weight attached to the moral goal the degree of morality. A special case is the familiar homo oeconomicus, who attaches zero weight to morality. At the other extreme one finds homo kantiensis who attaches unit weight to morality* (Alger and Weibull 2013, 4). Wang (2015, 56) takes a similar stance: *"The fact is, in actual economic activities, that the subject thereof is the "moral man", bringing with him economic, social, and environmental responsibilities, whose content is much richer than the "economic man." Therefore, in terms of production, man is certainly controlled by certain consciousness and guided by certain values. The moral consciousness of man directly affects and restricts his enthusiasm and energy release."*

What the current paper is adding to the homo moralis notion is locating and describing its selflessness dimension. About two and a half centuries ago, Adam Smith (1759, 1) wrote: *"No matter how selfish you think man is, it's obvious that there are some principles (…) in his nature that give him an interest in the welfare of others, and make their happiness necessary to him (…)"*. The human trait called rational selflessness in this paper is rational, because – ideally – humans would realize under certain circumstances (i.e. having enough information and knowledge) that using more resources would only lead to waste rather than increase well-being, so they would redistribute resources among community members. For multiple reasons this mostly does not happen. The general mindset sees the system as open, growth possibilities as infinite, and finally, needs as insatiable. Neither of these three are true in moral economics. We live in a closed system with limits, and the individuals' consumption can be stopped at a certain level while maintaining their well-being sustainably. However, once the latter recognitions reach collective consciousness, following self-interest (as per the *invisible hand*) will be accompanied by a certain self-control.

In fact, the notion of *the invisible hand* can also be interpreted in the moral context, as Wang (2015, 57) did it: *"As an invisible "hand of rationality" or "power of rationality", morality realizes the rational operation of all kinds of capital involved in production, guiding people to maximize profits."* Parallels may also be drawn between rational selflessness and the *objective self-interest* defined by the Hungarian economist Laura Baritz: *"Real human needs and interests incorporate wanting others' to fare well."*[6] And putting it in the negative way, as Wright (1999) quotes Herbert Spencer: *"No one can be perfectly happy till all are happy."*

Rational selflessness as a mechanism has two theoretical proofs: One is that humans cannot reach the top of their needs pyramids until others have reached similar levels of well-being, because the *"respect for others"* need – leading optimally to moral self-justification – would not be satisfied. The other is that humankind cannot climb higher on the progress pyramid without increasing cooperation. The two phenomena are linked, as cooperation lifts others up. This is one genius of human design, and may be a reason for the *"link between human nature and human history"* (Wright 1999) in the current paper's interpretation, namely, that the needs pyramids and the progress mountain are linked.

5.4.2 The Sharing Multiplier

This chapter will aim to describe how sharing multiplies the benefits gained from sharing value. The value shared could come from multiple sources: from sharing information, knowledge (know-how or know-why), (best) practices, certain services (e.g. entertainment-related), or even physical property (flats, cars, laptops, etc.). Examples of non-shareable goods include food and water, but through organizing, wasting any of it can be avoided.

For the sake of simplicity, let the value of the shared good be 1. Every person in the model owns something of value 1. One connection or one share between two people also has the value of 1, as derived from rational selflessness. The number of people is a positive integer: p. If we share something, we are creating a higher value than that of what we share. As with sharing a psychological need of sharing or helping is satisfied, the value increases by the number of the connections between people ($[p \times (p-1)]/2$). If they are all connected to each other, and if they all share what they own ($p \times 1$), the value formula looks like this:

$$p + ([p \times (p-1)]/2)$$

Simplified:

$$(p^2 + p)/2$$

This is $(p+1)/2$ times more than p (the number of people) itself, which was the initial value of the to-be-shared good, as everyone had 1. So if we want to create extra value through sharing, p – i.e. the number of people willing to share value – needs to be at least 2, because that is when the multiplier is bigger than 1:

$$(p+1)/2 > 1$$

[6] Tóth quotes Baritz (2014, 59). Own translation. Original in Hungarian: *"Az ember valódi szükségletében, önérdekében a mások jóllétének akarása is benne van."*

5.4.3 *The Basic Equation*

The difficulty of writing moral economic equations lies in the fact that they aim for creating and describing an economically perfect world, while the world has never been perfect as far as we know.

Let us use the following signal letters to describe the moral macro-equilibrium:

$$p = \text{population number}$$
$$O = \text{potential output}\left(\text{positive}\right)\text{of human activities}\left(= f\left(p\right)\right)$$

N = needs (=f(p)); In reality, people have several different needs, but these will be seen as unified in this chapter, as the complexity of measuring needs exceeds the scope of the paper.

Ideally, i.e. if the systems of distribution worked perfectly, humans only needed to produce as much output as is required to satisfy their needs. Thus the following equation can be written:

$$O = N$$

Both sides of the equation are population dependent, so this equation defines the ideal population number (at a given level of technology). An estimation of how these functions would look like and how they would intersect can be seen in Diagram 5.1. At a given level of technology (ceteris paribus), it is assumed that the potential output equals

$$p + ([p \times (p-1)/2)$$

where p is the number of people, and (p × (p – 1)/2) the number of their connections (best case scenario – they are all connected). This function represents the value people create by themselves and by sharing.

As is visible on the diagram, an intersection point exists. At a population number below that point potential output is less than the needs to be satisfied. When translated into a realistic economic scenario, this means scarcity. Exceeding the ideal population however also leads to economic problems: unemployment, waste of resources, and an economic crisis requiring artificial demand boosts.

As this diagram stands for a given level of technology, this is also the main determinant of and the main limit to the ideal population number. When writing "level of technology", each level can be understood as a stage of materialized cooperation, including connectedness, with an ever-improving infrastructure. Levelling up in technology terms requires historic discoveries and inventions, e.g. from the past: fire lighting, the alphabet, the compass, the printing press, the steam engine, the telegraph, etc.

The model assumes that no output is wasted. This requires us to consume everything we produce. Recycling and renewable energies are thus crucial. The human

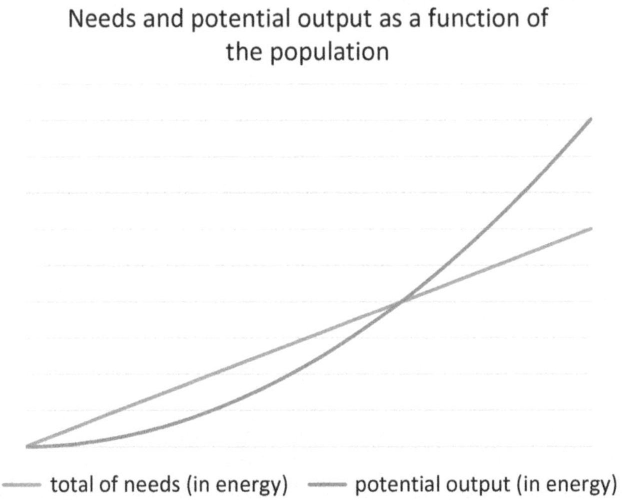

Diagram 5.1 Total of needs and potential output as a function of the population. (The diagram was established by the author)

input into these industries has to be incorporated into the calculation of potential output, so that all human potential is realized while producing exactly as much as is needed. On the other side of the equation, punctual and up-to-date need-quantification is just as important, as already pointed out in the paper.

The diagram is only for the illustration of the relationships' (between population number, needs and output) nature, as it is beyond the scope for this paper to calculate the needs and potential output of billions of people, who are not even fully connected, as of now. Also, the needs and the output functions are currently more sensitive in reality, as there are several influencing factors. However, in theory, this (Diagram 5.1) is what aggregate needs and output look like in the moral economy. Thus it is not the numbers, and not even the ratios on the diagram that are realistic and important, but the form of the needs and the potential output functions.

Factors that diverted attention from this rather evident relationship, and that hid the form of the needs and output functions, were: the lack of connectedness, the dependence on non-renewable energy resources and the belief of human needs being infinite and insatiable.

5.4.4 *The Moral Economic Measurement of Inequalities*

The following micro-indicator of task sharing's inequality can be applied to labour division, if the tasks are measurable, despite the differences in the nature of tasks. Let us assume that two people (who represent the micro-level) are supposed to share a task. The basic indicator of the inequality of task sharing between two people can be calculated as follows:

$$U = \left(\|I_1 - I_2\|\right) / \left(I_1 + I_2\right)$$

With words: The unfairness factor equals the absolute value of the difference between the individual commitments, divided by the sum of the input of the individuals.

As the absolute value of the difference is in the nominator, the lower the value of the whole indicator, the better (the less unfair the task sharing). The lowest possible value is 0, the highest 1. If the value is 1, that means one of the individuals is completely relying on and exploiting the other.

For the distribution of wealth between two people, an equation of a similar structure can be used:

$$U = \left(\|W_1 - W_2\|\right) / \left(W_1 + W_2\right)$$

According to the concept of rational selflessness, inequality of wealth (or income) affects well-being negatively. Analogously, inequality of task sharing affects productivity negatively.

On the micro-level, both the level of well-being and that of productivity have to be multiplied by (2-U) in order to get a result. If U is 0, that gives the potential well-being or productivity, without inequality, i.e. the double of the assumed basis. The Swedish proverb *"Shared joy is double joy"* comes to the mind.

An illustration of task sharing can be found on the diagram below. The "Result" curve, which has the value (2–U) × 100, mirrors the moral economic fact that sharing has added value, manifesting itself in positive externalities (Diagram 5.2).

On the macro level, it is similar to say that a country's GDP would have the potential to be: the given GDP + the given GDP multiplied by (1 – the Gini coefficient).

Potential means that this would be the given country's GDP, if there was an equality in distribution and sharing tasks. To illustrate this, the author extracted GDP/capita data for the 80 countries where the Gini index was available for the year of 2010. (The reason that the year 2010 was selected is that this is the latest round year with the most available data.) The numbers have been sorted by GDP/capita ascendingly. As is visible from the diagram, equality would make the biggest efficiency difference in the relatively wealthiest of the countries, some of them with relatively wide income gaps (Diagram 5.3).

The prevalence of rational selflessness however (which is a requirement to achieve the full potential of a given community) depends on the level of interconnectedness. Interconnectedness in a given community can be measured by the number of existing connections (or opportunities to connect), divided by the number of all possible connections: $r/((p \times (p - 1))/2)$. If this ratio reaches 1, the community is fully connected. The more interconnected a community, the higher there are the

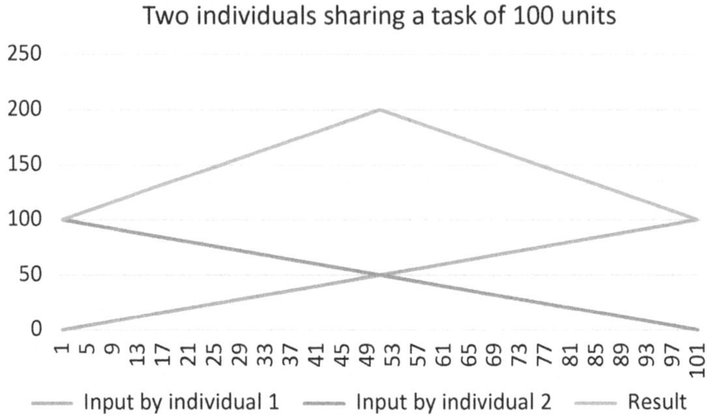

Diagram 5.2 Task sharing and its efficiency according to the laws of moral economics. (Established by the author)

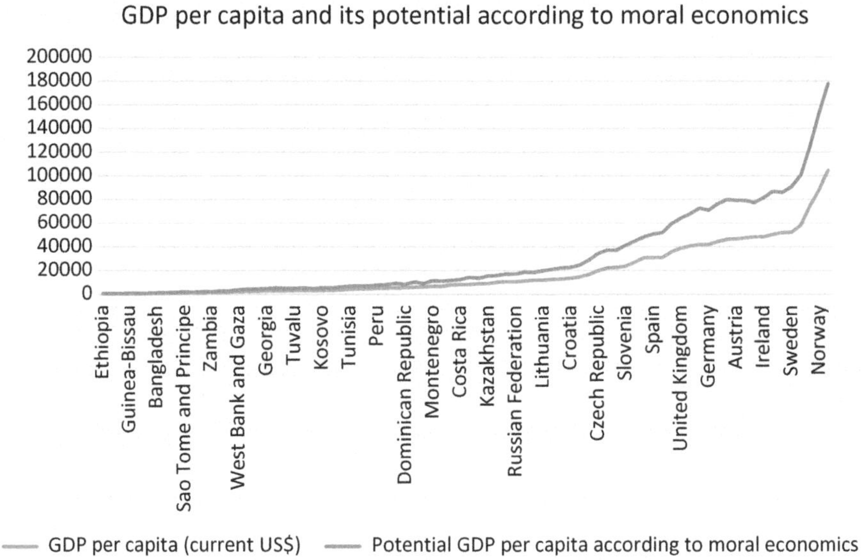

Diagram 5.3 GDP per capita and its potential according to moral economics. (World Bank 2010; and edits by the author)

chances for rational selflessness to create mechanisms driving the community towards fairness, thus also towards higher levels of productivity and well-being. This sub-conclusion, and the equations presented in this chapter together are a starting attempt to give the (moral) economic justification of equality.

5.5 Tools and Solutions

This chapter is about relating certain phenomena in the economy to moral economics, because they can function as tools to reach a higher state of well-being.

5.5.1 Robotization and Universal Basic Income

We have a quick and massive robotization process on our threshold, which promises to lead to abundance, yet threatens with the devaluation of many human activities and jobs. Many believe that this will be the first era, where we can eliminate the human factor from the production process.

Meanwhile, the policy of Universal Basic Income is emerging as a potential solution to present-day economic problems. *"Governments around the world are evaluating its use, and some are embarking on pilot studies"* (Coppola 2017). Some would call UBI the social vaccine of the twenty-first century. Their approach has similar characteristics with the hope of fully automated luxury communism (Sadowski 2016).

Robotization trends are interlinked with Universal Basic Income proposals. This is perfectly illustrated by the 2015 EU *Draft Report with recommendations to the Commission on Civil Law Rules on Robotics* (Delvaux 2015). The report came to public attention mostly through introducing the concept of electronic persons (Delvaux 2015, 12) but it refers to the new industrial revolution, likely to be unleashed by *"sophisticated robots, bots, androids and other manifestations of artificial intelligence"*, and also expresses the *"concerns about the future of employment"* (Delvaux 2015, 3). It considers the cases of autonomous vehicles, medical robots, and human repair and enhancement, just to name a few examples (Delvaux 2015, 8–9). Up until now, this plan is the most comprehensive and detailed public legal document on the two, linked issues of robotization and Universal Basic Income. The comprehensiveness lies in the suggestion of *"a coherent approach to regulation at European level"* (Delvaux 2015, 4) and in inviting all Member States to seriously consider a general basic income in the light of the possible effects on the labour market of robotics and AI (Delvaux 2015, 10).

Moral economics supports robotization, as it drives us towards sustainability. The challenges robotization poses to the labour market (i.e. the challenge of unemployment) can be neutralized by the many more potential channels, connecting people in need to those with the solutions. However, the system may indeed need buffers, such as Universal Basic Income. Critics could ask how motivation to work would be sustained. The response is that meaningful work is a human need as well.

5.5.2 *Blockchain in the Moral Economy*

The moral economy could draw on blockchain technology, mostly because of the accountability aspect arising from the potential use of smart contracts, but also because of yet undeveloped implementation ideas and opportunities, such as a common blockchain calendar. According to The Economist, blockchains are *"a way of making and preserving truths"* (The Economist 2015). The moral economic implication of a blockchain calendar is that important actions and events are *"set in cryptographic stone"* (The Economist 2015) enforcing honesty and fairness, both characteristics of the moral behaviour set.

Accountability contributes to order as an important feature of orderly social systems. Blockchain provides the technological – and to a certain extent even the legal – infrastructure for accountability and transparency. Some developers in the so-called "Crypto Valley" (Zug, Switzerland) are already sketching applications of blockchain, in which all state expenditures, i.e. the whole government budget is traceable. The implementation of these "social smart contracts" is said to be an ultimate game-changer in the operations of politics ((Mohácsi 2018).

But as the paper shall stay more or less within the boundaries of economics, the focus will now be turned on the moral economic aspects of implementing blockchain technology in order to reach an accountability coverage, meaning that for all possible event outcomes (especially those negatively affecting humans) somebody can be held responsible. Moral economics is designed for sharing, and this raises a privacy concern, at least as long as the privacy-transparency gap has not been bridged. Through accountability and its legal consequences, the checks and balances of the legal framework to sharing could be planted. This would prevent the abuse of information on the sites and applications through which sharing is organized. The latter statement's significance is shown by a tendency: the currency of value in certain fields, especially online, is shifting from money, through personal information, to trustworthiness.[7] Although, as Sedláček (2012, 117) would argue: money itself, as a social abstraction, is based on trust, an unwritten social contract, detached from matter, space, or even time (2012, 122).

Also, there is word about an "Internet of Value" (Mohácsi 2018). This "Value Web" would make value exchange *"as easy as exchanging information today on the web"* (Larsen 2015). Furthermore, the author of this paper envisions a blockchain based calendar, where people could share and verify events, actions and plans throughout time. And there exists the institution of smart contracts. The paper will not go into detail regarding smart contracts, but one feature of them needs to be mentioned: They are deterministic. They cover all possible scenarios and enable a complete scenario analysis (Mohácsi 2018). This characteristic combined with the blockchain calendar could ultimately lead to predicting humankind's future together.

[7] For more information in this topic, see the 2012 TED Talk by Rachel Botsman: *The currency of the new economy is trust.* https://www.ted.com/talks/rachel_botsman_the_currency_of_the_new_economy_is_trust/transcript?nolanguage=eg (Accessed March 28th, 2018).

In one sense, blockchains are trusted third parties (The Economist 2015) and in a more subtle, moral one, they fulfil the role of impartial spectators, with enforcing morality through being *"truth machines"* (The Economist 2015).

5.5.3 Artificial Intelligence and the Internet of Minds

> When you look beneath the roiled surface of human events, beyond the comings and goings of particular regimes, beyond the lives and deaths of the "great men" who have strutted on the stage of history, you see an arrow beginning tens of thousands of years ago and continuing to the present. And, looking ahead, you see where it is pointing. (Wright 1999)

Channelling creates a network, and the most comprehensive network known today is the Internet, not just in itself, but interweaving our infrastructure in many ways (for example as the Internet of Things). An enhancing technology to this network is artificial intelligence. AI is currently an *"aspirational term reflecting a goal"* (Pavlus 2017). This goal is to be achieved by making progress in machine learning, deep learning and developing neural networks (Pavlus 2017). Much of our image about AI and its future is speculation. In general, opponents of AI and robotization view the opportunity of singularity with scepticism. They imagine a world, where nature and human life are completely simulated, all data is stored, retrievable and evaluated, and machines can answer all questions and remember everything that has ever happened. They fear the loss of basic human characteristics and peculiarities (MAK 2017).

This view, generalized, resembles the furthest look into the future. On the mid- and short term, we find scientists, and leaders of giant technology companies worrying about artificial intelligence, for example: the late physicist and cosmologist Stephen Hawking, Bill Gates and Elon Musk (Balkam 2015). For part of the general public, the spread of artificial intelligence and robotization gives the impression of a threat to privacy, jobs and potentially their safety. However, we might have several reasons to change this perspective and to embrace AI. It is already made use of in combatting infectious diseases, tackling gun violence, fighting cancer and sight loss, and managing energy supply – just to cite a few examples (Gray 2017). Changing to an AI interwoven system is sometimes compared to other events in the history of technological development. *"AI will enliven inert objects in the way electricity did over 100 years ago"* (Balkam 2015). Thus, *"We should view AI not as something competing with us, but as something that can amplify our own capabilities"* (Gray 2017).

Given the benefits from AI for the healthcare and security sectors, AI could also help humans fill economic knowledge and information gaps in order to satisfy all needs, by – in general terms – filling the spaces and connecting the dots of the economic system. Early movers have already spotted this opportunity, for example in the real estate industry (Sicklick 2017).

The most developed and feasible network we can imagine at present is the Internet of Minds. This network would connect and give access to each other's thoughts. Implementing this network, potentially enhanced with artificial intelligence, would most probably lead to opportunities not even known as of today to humankind. There have been recent events which mark the first milestones of the process. Last September, an article emerged stating in the title that *"Researchers Have Linked a Human Brain to the Internet for the First Time Ever"* (Caughill 2017). The move turned the brain into a node in the system. For now, the experiment was one directional, but *"In future, there could be information transferred in both directions – inputs and outputs to the brain"* (Caughill 2017).

The internet of minds would be a new means for transparency, and a new type of organism. Its creation would be a moral revolution in itself. The Finnish computer scientist Harri Valpola has already started laying down the technological basics for an internet of minds. From the perspective of this paper, one of Valpola's statements has unique importance: *"In the first wave of AI you had to be a programmer. In the second wave of AI you have to be a data scientist. The third wave of AI – the more moral you are the better"* (Manthorpe 2017).

In the current interpretation, morality in the field of AI is not only needed in order to create moral AI beings. Artificial intelligence – whatever its form may be – would soon recognize the genius design of humankind, where morality is the crucial strategic element of sustainability. Thus, humans would also need to be moral in order to cooperate with AI, a more knowledgeable entity or entities than humans themselves, for their own good.

The previous statements, of course, only hold true if morality is naturally a positive function of intelligence, especially from the point on when human capabilities are surpassed by AI. However, this paper assumes that, similarly to the other five axioms laid down at the beginning.

5.6 Outlook (Into Space)

Elon Musk once said: *"I would like to die on Mars, just not on impact"* (Al-Greene 2013). If moral economic tendencies are projected on the more distant future, the renowned tech entrepreneur proves to have reached a very logical conclusion.

Further improvements in technology would demand lower and lower population numbers, according to the basic equation. The reason for that: the decreasing intersection point of total needs and potential output. If we use the level of technology as a multiplier of potential output, the quadratic function "draws closer" to zero, with every improvement. With rising technology levels, fewer and fewer people are needed to sustain themselves while their needs are staying satisfied. Following this logic would lead to a diminishing population and the end of humans as a species.

An exciting way not to become extinct is to find new purposes, such as inhabiting other planets. This would lead to *"Making Humans a Multi-Planetary Species"* (Musk 2017) (Diagram 5.4).

Diagram 5.4 Intersections of needs and potential output at different levels of technology. (Own design)

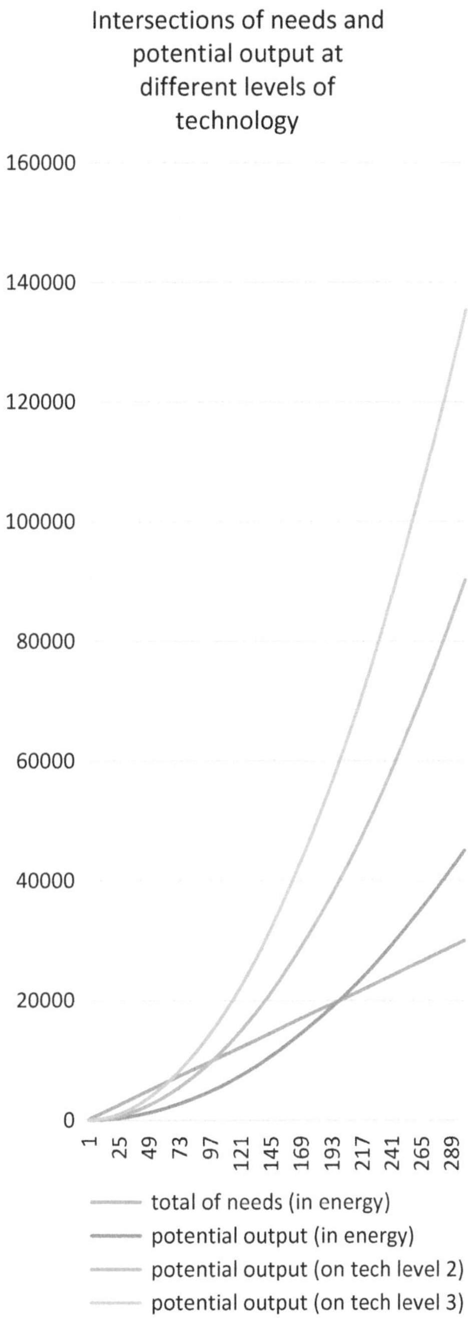

5.7 Conclusion

Part of the physicist profession is still on the quest for a unified theory, and so are certain economists in their own field. The author of this paper believes that regarding economic theory, the unified theory must lie in moral economics, or that at least it is the clearest path thereto. The conclusion of this paper is aimed at supporting the previous statement, through drawing the very essence from the paper and through presenting the consequences.

The paper started with formulating its purpose, and introducing the axioms which moral economics stands on: that morality is objective, further, that humankind constitutes one big community whose goal it is to increase their well-being, that human motivations include rational selflessness, and that human needs are neither infinite, nor insatiable. Upon these axioms the paper has started to introduce a new school of economics.

A whole chapter was dedicated to being clear about the meaning and the ideological position of the moral economic concept. The usage of the words "moral economy" and the apparently newly coined term "moral economics" are reflecting in this paper a different approach from any earlier attempt, in their comprehensiveness, their strength through the arguments and models, and their completeness. It was not mentioned in the ideological chapter yet, but it can be drawn as a conclusion from subsequent chapters that – in a rather political context – the moral economy unites the freedom of a bottom-up with the security of a top-down society.

The reflection on Smithian economic theory has shown how we are witnessing today, in real-time, the proliferated successes and malfunctions of the mechanisms Smith described over two centuries ago. However, present-day capitalisms are different from "pure" Smithian capitalism. The systems contain morality. Several firms and owners of capital have recognised that fostering and cultivating corporate morality pays off on the mid- and long term. Furthermore, a top-down perspective is starting to make more sense than continuing with bottom-up mechanisms, at least over the limits of sustainability, i.e. in poverty alleviation and environment protection.

Models were formed where economics and ethics could coherently and logically be fitted into. Economic value can be stabilized through adjusting the factors influencing aggregate demand and supply (needs and output), and striking a balance between them. Economic order (or orderliness) is to be taken more seriously and to be acted upon with more responsibility, on the level of a closed system, i.e. incorporating our natural environment into the economy.

At the heart of moral economics lies rational selflessness, a relatively undiscovered side of economic behaviour and an inherent attribute of humans. Rational selflessness was discovered by the author as a hidden building block of the Maslowian (Maslow 1943) needs pyramid, and then made it far in the theory of this paper as a determinative human feature, as well as a motivational concept. The connections between the models, leading to find the intersection point of the needs and output functions, were established through rational selflessness.

The multiplicator effect of sharing value and tasks contributes largely to moral economics' feasibility. The sharing multiplier is a combinatorial one, on which the output function is based, but its underlying philosophy has an effect on sharing from even between as few as two people, up to nations' economies.

Creating order through channelling needs and their responses has power. Through making motivation and information meet, it brings humankind to a shift, for which it is more than ready. The shift involves the implementation and maintenance of artificial intelligence, applying the policy of Universal Basic Income, implementing blockchain technology broadly and creating a transparent environment, amongst other measures. Regarding blockchain, the paper projects the invention of a blockchain calendar which would ultimately extend the accountability coverage, thus creating a "blockchain of good." The Internet of Minds was also envisioned.

The mechanism of the intersection point (that of needs and potential output) drawing closer to zero, and thus human work and input becoming less needed, can be compensated with setting further goals, such as inhabiting other planets. This, of course, has second priority at the moment to creating order in the current system. However, the preparations for the more distant future can already be a part of the current order.

With regard to further research opportunities, there is both space for research in the theoretical and in practical directions.

In the theory, the exact concept of economic value in the moral economic context requires further exploration, both in narrower and broader interpretations. Also, the concepts of rational selflessness and order could be deepened, in the philosophical sense, meaning that they should be deconstructed first. Questions arise where they came from or how they came into being, and whether the answers to these questions contribute to the philosophy of moral economics.

The moral economic stance on fair distribution should be clarified, answering the questions when and to what extent there should be equality, and on what basis, other than the human needs. As the models have shown, there might be more justification to moral economics than to a simple egalitarian utopia. Moral economics does not define concrete taxes or concrete ratios for fair distribution. Equality is aimed for on the very long (secular) term.

The practical research possibilities all relate to the implementation of moral economics. Firstly, the equation of aggregate needs and potential output should be translated into real numbers, through quantifying the level of technology as well. Second, the exact mechanism of channelling and how it makes improvements ought to be explored and be described with examples, in order to justify more-than-zero-sumness. As a value and moral economic tool, transparency should be incorporated, which largely contributes to channelling. Morality should also be studied on the level of institutions, organizations and companies.

After reaching the national level in moral economic studies, the research will not be far from establishing country specific implementation plans, including assessments of potential negative consequences, if the measures were applied too early. There should be an emphasis on bringing different moral systems closer together, and on learning from each other. From the Chinese perspective, for example, it seems that *"the most developed countries in the world (…) they have never really understood the special roles played by morality in the economy"* (Wang 2015, 55). This shows that global knowledge about morality still needs to be integrated. To quote Wang (2015, 72) again: *"… moral capital realizes its value through its own value progress. On the one hand, such progress mainly lies in the improvement of various expressly stated moral norm systems, and the rationality and feasibility of expressly stated moral regulations and rules. On the other hand, it lies in the constant assimilation between not expressly stated moral spirits, moral beliefs and moral concepts".*

References

Adami, C., and A. Hintze. 2013. Evolutionary instability of zero-determinant strategies demonstrates that winning is not everything. *Nature Communications*. https://www.nature.com/articles/ncomms3193.pdf. Accessed 22 Dec 2017.

Alger, I., and J.W. Weibull. 2013. Homo moralis – Preference evolution under incomplete information and assortative matching. *Econometrica* 81 (6). https://www.tse-fr.eu/sites/default/files/medias/stories/sem_12_13/eco_theo/weibull.pdf. Accessed 9 Oct 2017.

Baritz, L. 2014. *Háromdimenziós gazdaság – Lehet-e gazdálkodni erényetikai paradigmában? (Three-dimensional economy – Is it possible to do business in the virtue paradigm?)*. Doctoral dissertation, Corvinus University of Budapest, PhD-program.

Götz, N. 2015. 'Moral economy': Its conceptual history and analytical prospects. *Journal of Global Ethics* 11 (2): 147–162. http://www.tandfonline.com/doi/full/10.1080/17449626.2015.1054556. Accessed 6 Nov 2017.

Kolodko, G.W. 2018. Socialism or capitalism? Tertium Datur. Conference paper presented at the workshop: *Comparative economics and economic systems – The impact of János Kornai's thoughts*. 18 January, 2018, Budapest. http://www.kornai90.hu/wp-content/uploads/2018/01/KOLODKO_Tertium-Datur_ENG.pdf. Accessed 11 Apr 2018.

Maslow, A.H. 1943. A theory of human motivation. *Psychological Review* 50: 370–396. http://psychclassics.yorku.ca/Maslow/motivation.htm. Accessed 4 Aug 2017.

Musk, E. 2017. Making humans a multi-planetary species. *New Space* 5 (2). http://online.liebertpub.com/doi/pdfplus/10.1089/space.2017.29009.emu. Accessed 12 Sept 2017.

Piketty, T. 2014. *Capital in the twenty-first century*. Cambridge, MA: Belknap, Harvard University Press.

Sayer, A. 2000. Moral economy and political economy. In *Studies in Political Economy* 61, Spring 2000. http://spe.library.utoronto.ca/index.php/spe/article/viewFile/6751/3748. Accessed 20 Dec 2017.

Sedláček, T. 2012. *A jó és a rossz közgazdaságtana. A Gilgames-eposztól a Wall Streetig* (Economics of good and evil. The quest for economic meaning from Gilgamesh to Wall Street). Budapest: HVG Publisher. (Hungarian edition).

Smith, A. 1759. *The theory of moral sentiments*. Online version: http://www.earlymoderntexts.com/assets/pdfs/smith1759.pdf. Accessed 14 July 2017.

———. 1776. *An inquiry into the nature and causes of the wealth of nations*. https://www.ibiblio.org/ml/libri/s/SmithA_WealthNations_p.pdf. Accessed 21 Dec 2017.

Tóth, G. 2016. *Gazdaséggép* (Economic machine). Budapest: L'Harmattan Kiadó.

Varoufakis, Y. 1998. *Foundations of economics: A beginner's companion*. London: Routledge.

Wang, X. 2015. *On moral capital*. Berlin/Heidelberg: Yilin Press, Ltd/Springer.

Online Sources

Al-Greene, B. 2013. Elon Musk at SXSW: 'I would like to die on Mars, just not on impact'. *Mashable*, March 10. http://mashable.com/2013/03/10/elon-musk-sxsw-2013-mars/#Kj893vyC8qqC. Accessed 1 Aug 2017.

Balkam, S. 2015. What will happen when the internet of things becomes artificially intelligent? *The Guardian*, February 20. https://www.theguardian.com/technology/2015/feb/20/internet-of-things-artificially-intelligent-stephen-hawking-spike-jonze. Accessed 8 Oct 2017.

Caughill, P. 2017. Researchers have linked a human brain to the internet for the first time ever. *Futurism*, September 14. https://futurism.com/researchers-have-linked-a-human-brain-to-the-internet-for-the-first-time-ever/. Accessed 8 Oct 2017.

Coppola, F. 2017. The IMF gives a cautious welcome to universal basic income. *Forbes*, October 15. https://www.forbes.com/sites/francescoppola/2017/10/15/the-imf-gives-a-cautious-welcome-to-universal-basic-income/#51e5c37c2b64. Accessed 22 Oct 2017.

Delvaux, M. 2015. Draft report with recommendations to the Commission on Civil Law Rules on Robotics (2015/2103(INL)). Committee on Legal Affairs. http://www.europarl.europa.eu/sides/getDoc.do?pubRef=-//EP//NONSGML%2BCOMPARL%2BPE-582.443%2B01%2BDOC%2BPDF%2BV0//EN. Accessed 27 Mar 2018.

The Economist. 2015. The great chain of being sure about things. October 31. https://www.economist.com/news/briefing/21677228-technology-behind-bitcoin-lets-people-who-do-not-know-or-trust-each-other-build-dependable. Accessed 31 Mar 2018.

Gray, R. 2017. Treating cancer, stopping violence… How AI protects us. *BBC Future*, September 15. http://www.bbc.com/future/story/20170914-spotting-cancer-stopping-shootings-how-ai-protects-us. Accessed 8 Oct 2017.

Larsen, C. 2015. Toward an internet of value: An interview with Chris Larsen, CEO of Ripple Labs. https://www.mckinsey.com/~/media/McKinsey/Industries/Financial%20Services/Our%20Insights/Toward%20an%20internet%20of%20value/Toward%20an%20internet%20of%20value.ashx. Accessed 28 Mar 2018.

Manthorpe, R. 2017. Harri Valpola dreams of an internet of beautiful AI minds. *Wired Magazine*, September 23. http://www.wired.co.uk/article/harri-valpola-curious-ai-artificial-intelligence-third-wave. Accessed 8 Oct 2017.

Pavlus, J. 2017. Stop pretending you really know what AI is and read this instead. *Quartz*, September 6. https://qz.com/1067123/stop-pretending-you-really-know-what-ai-is-and-read-this-instead/. Accessed 12 Sept 2017.

Sadowski, J. 2016. Why Silicon Valley is embracing universal basic income. *The Guardian*, June 22. https://www.theguardian.com/technology/2016/jun/22/silicon-valley-universal-basic-income-y-combinator. Accessed 3 Sept 2017.

Sandmo, A. 2014. Adam Smith and modern economics. Norwegian School of Economics. https://brage.bibsys.no/xmlui/bitstream/handle/11250/194583/SAM1314.pdf?sequence=1. Accessed 25 Dec 2017.

Sicklick, J. 2017. Artificial intelligence will remake the real estate industry – For the better. *LinkedIn Pulse*, September 14. https://www.linkedin.com/pulse/artificial-intelligence-remake-real-estate-industry-better-sicklick/. Accessed 8 Oct 2017.

Stanford Encyclopedia of Philosophy. 2013. Adam Smith's moral and political philosophy. https://plato.stanford.edu/entries/smith-moral-political/. Accessed 21 Dec 2017.

Wright, R. 1999. *Nonzero – The logic of human destiny*. http://www.nonzero.org/toc.htm. Accessed 29 July 2017.

Zuckerberg, M. 2017. Building global community. *Facebook note*, February 16. https://www.facebook.com/zuck/posts/10154544292806634. Accessed 9 July 2017.

Databases

The World Bank: *GDP per capita (2010, current US$).* https://data.worldbank.org/indicator/NY.GDP.PCAP.CD. Accessed 29 Oct 2017.
The World Bank: *GINI index (2010, World Bank estimate).* https://data.worldbank.org/indicator/SI.POV.GINI Accessed 29 Oct 2017.

Other Sources

BBC World Service. 2017. Adam Smith: Father of capitalism. *The Forum (radio show)*, November 22. http://www.bbc.co.uk/programmes/w3csvsfb. Accessed 21 Dec 2017.
MAK – Österreichisches Museum für angwandte Kunst – Museum of Applied Arts. 2017. Artificial tears – Singularity and humanness – A speculation. *Exhibition brochure.*
Mohácsi, Z. 2018. *Blockchain: A digitális bizalom alapja? / Okosszerződések* (Blockchain – The currency of digital trust? / Smart contracts) March 22. Presentation series by the Cybereconomics Excellence Center of the Budapest Business School – University of Applied Sciences.

Chapter 6
How (Not) to Connect Ethics and Economics: Epistemological and Metaethical Problems for the Perfectly Competitive Market

Caspar Willem Safarlou

Abstract This paper addresses Joseph Heath's attempt to derive moral obligations from the conditions that are specified by the model of the perfectly competitive market. Through his market failures approach to business ethics he argues that firms should behave as if they are operating in a perfectly competitive market. However, I argue that this derivation of moral obligations runs counter to the metaethical principle that moral actions need to be voluntarily chosen from a set of alternatives. To the extent that Milton Friedman's derivation follows the same lines, my objection is also applicable to his approach to business ethics. I bring out the fact that the conditions required by the model of the perfectly competitive market cannot be realized in the actual world and argue that this causes problems for any moral obligations that might follow. My objection is illustrated by an intuitive example of someone set to an impossible task. I also bring in a way that Heath could work around this objection, but I argue that this would imply the collapse of his approach into another kind of theory that he wishes to distinguish himself from. More deeply, I show that my metaethical objection has epistemological consequences that undermine the very basis of the model of the perfectly competitive market. I conclude by stating that we need a different conception of competition, pointing to the facts that such a perspective would need to take into account, and suggesting that the concept of rivalry is up to the job.

6.1 Introduction

In recent years, Joseph Heath has provided a novel moral defence of markets with his market failures approach to business ethics (Heath 2014, vii-viii; 173; 199n14). His approach holds that firms have a moral obligation not to "seek to profit from

C. W. Safarlou (✉)
Utrecht University, Utrecht, Netherlands

P. Róna et al. (eds.), *Words, Objects and Events in Economics*, Virtues and Economics 6, https://doi.org/10.1007/978-3-030-52673-3_6

market failure" on the basis of the economic model of the perfectly competitive market (Heath 2014, 89). This basis uniquely distinguishes Heath's approach from other stockholder theories that attempt to derive moral obligations by also involving general morality.[1]

I share Heath's goal of providing a moral defence of the free market, but I do not think that the model of the perfectly competitive market is suited to this end. My primary aim here is to show that his attempt to derive moral obligations from the model of the perfectly competitive market contradicts the metaethical principle that moral actions need to be voluntarily chosen.

Although I think that this metaethical problem originates in the deeper epistemological foundations of the model of the perfectly competitive market, the brunt of my criticism is directed towards the metaethical component of Heath's justification of the market failures approach. Without first trying to make strong claims about these deeper foundations, I will show that the issues that I will tackle are important for any usage of the model of the perfectly competitive market in an ethical context. For example, I will also show how Milton Friedman's business ethics is vulnerable to the same line of criticism. My analysis of Heath's justification of the market failures approach will thus centre on the way in which moral obligations are derived from the model of the perfectly competitive market.

First, I will start out by explaining the way in which Heath tries to derive moral obligations from the model of the perfectly competitive market. This will include the way in which he relates the role of the government to the privileges and operation of firms and the role that he assigns to the model of the perfectly competitive market. I will also touch upon the way in which Friedman's business ethics tries to perform the same kind of derivation of moral obligations (as interpreted by Heath).

Second, I will provide a metaethical objection to Heath's (and by extension, Friedman's) attempt at deriving moral obligations from the model of the perfectly competitive market. I bring out the fact that the conditions required by said model cannot be realized in the real world and argue that this causes problems for any moral obligations that might follow. This objection is illustrated by an intuitive example of someone set to an impossible task. I also bring in a way that Heath could work around this objection, but I argue that this would mean the collapse of his approach into another kind of theory that he wishes to distinguish himself from. I conclude this section by pointing out why it is in principle impossible to provide such a workaround, as my metaethical objection shows a deeper fault within the epistemological basis of the model of the perfectly competitive market.

Lastly, I conclude that we need a different conception of competition for correctly theorizing about competition from a moral and epistemological point of view and point to a conception of competition that seems to be a good alternative.

[1] For a comparison between Heath's foundational approach and those of Langtry and Goodpaster, see Heath (2014, 90).

6.2 Heath's Market Failures Approach

Heath starts arguing for his market failures approach by observing that firms are legal constructs which have certain privileges regular citizens do not have (such as limited liability). This means that firms exist only by virtue of the fact that the government grants them these privileges. Consequently, Heath says, the government may "impose certain obligations [upon firms], in return for the privileges granted" (Heath 2014, 29). In principle, these obligations can range from ethical obligations that ask the firms to adhere to a certain set of rules to government-enforced regulations that are backed up by legal sanctions.

Next, Heath asks the fundamental question why governments should grant these privileges in the first place. His answer to this question is fairly complex, but it boils down to the idea that "society wants to encourage competition between suppliers" because it "secures the operation of the price mechanism" without which "you simply cannot organize a complex economy" (Heath 2014, 29; 30). Under the correct conditions, the price mechanism makes sure that the prices of goods are "cleared" – which means that there will be no unsold goods nor unsatisfied customers. According to Heath, we should look to the model of the perfectly competitive market to provide us with these conditions:

> *The central conclusion [of the first fundamental theory of welfare economics] is that the outcome of a perfectly competitive market economy will be Pareto-optimal—which means that it will not be possible to improve any one person's condition without worsening someone else's.* (Heath 2014, 29–30)

In a perfectly competitive market, there is a "race to the bottom" between suppliers of goods. Each supplier is able to achieve profit by lowering their prices under those of their competitors, making up for the lower price because they attain a larger number of sales. This means that competitors keep undercutting each other to the point at which all prices in the market are cleared and all profits have disappeared.[2] In the end, competition allows for "a more efficient allocation of [society's] resources and labor time" than economic systems in which competition is absent (Heath 2014, 30–31).[3]

In the next step, Heath connects this institutional argument back to the privileges that the government can grant to firms. This means that the conditions under which firms are to be granted their privileges are those dictated by the model of the perfectly competitive market. This provides the basis for both government regulation of firms and moral obligations for firms. Heath argues that firms need to be regulated by the government in such a way as to create the conditions that will make sure that the prices of the goods they produce will be cleared. Since profits are price signals that show whether there are still customer needs to be satisfied or resources to be put

[2] The technical name for these profits in the model of the perfectly competitive market is pure profits.

[3] Heath illustrates this contrast by discussing the way in which the absence of the price mechanism caused problems for the former Soviet Union Heath (2014, 30).

to a better use in another place, firms gain a moral obligation "to do what is necessary in order for the firm to maximize profits in this way" (Heath 2014, 31).

However, the conditions demanded by the model of the perfectly competitive market to achieve a Pareto-optimal outcome are not always met. In such a case, market outcomes are not Pareto-optimal and are called market 'failures' as they haven't lived up to the standard of the perfectly competitive market.[4] The next question that arises is: Should the state start regulating firms in order to make sure that market failures are prevented from happening?

Although Heath holds that there is a basis for the government to regulate firms through the legal mechanism because their privileges are granted by the government, he argues that these regulations are unfeasible. He says that the legal apparatus is "a somewhat blunt instrument" and that in too many cases "the state simply lacks the information needed to implement the necessary measures" that are needed to make sure that market failures do not occur (Heath 2014, 36–38). Even in cases where the government has enough information to regulate for situations in which the market fails, he argues that the administrative costs that are incurred in such a situation are so high that they turn government regulations into an unfeasible task. It is at this point in the argument that firms become subject to moral constraints that are not backed up by legalized force:

> *Imagine for a moment a deontically perfect world, in which everyone could be counted on to comply with all moral requirements. How should an ethical corporation behave in such a world? The answer is quite simple. The firm should behave as though market conditions were perfectly competitive, even though they may not in fact be.* (Heath 2014, 37)

In this respect, Heath agrees with Milton Friedman's metaethical approach to business ethics (Friedman 1962). In his interpretation of Friedman's business ethics, moral obligations of firms are also grounded in the model of the perfectly competitive market; firms are thus morally obligated not to exploit market failures (Heath 2014, 31–35). However, Heath is very critical of Friedman's approach and criticizes it on two grounds. His first criticism is directed at Friedman's derivation of normative ethics from his metaethical basis. According to Heath, Friedman "arbitrarily limits the set of obligations [for firms] to those that support only some of the many Pareto conditions [demanded by the model of the perfectly competitive market]" (Heath 2014, 35). In this respect, Heath wishes to be more consistent in the derivation of moral obligations from the model of the perfectly competitive market. He illustrates this through a very compelling argument that will at the same time provide us with a concrete example of the moral obligations that both authors have in mind:

> *...Friedman argues that pollution reduction is one of the illegitimate responsibilities pressed upon managers in the name of "social responsibility." But pollution is a negative externality—a cost associated with some economic activity that is transferred to a third party without compensation. These externalities exist because the set of markets is incomplete. We cannot exercise property rights over the air that we breathe, for example. As a*

[4] For a compelling critique of the very idea that markets can fail, see Simpson (2005).

> *result, while we can charge people for dumping noxious substances on land that we own, we cannot do the same when they dump it in the air. For this reason, one of the Pareto conditions effectively requires that there be no externalities. Any corporation that pollutes is essentially profiting from a market imperfection. This means that there is no difference, from the moral point of view, between deception and pollution—both represent impermissible profit-maximization strategies. Friedman's decision to prohibit deception, while giving the wink to environmental degradation, is arbitrary and unmotivated.* (Heath 2014, 35)

This brings us to Heath's second criticism of Friedman's approach, which is directed at the way in which Friedman thinks that we can approach the ideal of the perfectly competitive market. Let us take a closer look at what kind of ideal is presented. The abstraction of the perfectly competitive market is claimed to be an ideal model such as that of a frictionless plane or a mathematically perfect circle. Any attempt to approach such a type of ideal is a simple corollary of the ideal in question. For example, if one draws a circle on a whiteboard by means of a pair of compasses, then one approaches an ideal circle quite linearly and directly. Friedman phrases this idea as follows:

> *No one has ever seen a Euclidian line—which has zero width and depth— yet we all find it useful to regard many a Euclidian volume—such as a surveyor's string—as a Euclidian line. Similarly, there is no such thing as "pure" competition. Every producer has some effect, however tiny, on the price of the product he produces. The important issue for understanding and for policy is whether this effect is significant or can properly be neglected, as the surveyor can neglect the thickness of what he calls a "line."* (Friedman 1962, 120)

It is not this type of ideal abstraction that Heath finds problematic, but its function as a gauge for the approximation of the ideal itself. He notes that "we may be tempted to conclude that if perfect competition generates perfect efficiency, then near-perfect competition should generate as close as possible to perfect efficiency" (Heath 2014, 39). It is this line of reasoning that is blocked by the second-best theorem (Lipsey and Lancaster 1956):

> *This theorem shows that in a situation in which one of the Pareto conditions is violated, respect for all of the other Pareto conditions will generate an outcome that is less efficient than some other outcome that could be obtained by violating one or more of the remaining conditions. In other words, while perfect competition generates a perfectly efficient outcome, a situation that is as close as possible to perfect competition will not generate an outcome that is as close as possible to perfect efficiency.* (Heath 2014, 39)

According to Heath, the implication of this theorem is that Friedman is blocked from making "the big sweeping generalizations that were the stock-in-trade of economists of Friedman's generation" (Heath 2014, 40). If one wishes to approximate the ideal presented by the model of the perfectly competitive market, then one cannot use Friedman's top-down reasoning to achieve this end.

In order to derive actual moral obligations for firms, Heath wishes to use a more bottom-up approach.[5] He notes that every individual trade that takes place still causes a Pareto improvement and that this makes the Pareto-optimum of the

[5] For the reason why the second-best theorem can't be used to derive moral obligations, see Heath (2014, 40).

perfectly competitive market model irrelevant for everyday life. Instead of appealing to the Pareto-optimum, one would need to "appeal to the particular efficiency gains that the firm is able to realize among its shareholders, its employees, and its customers" (Heath 2014, 40). On the one hand, this means that the model of the perfectly competitive market is still the source and foundation of the moral obligations that firms have to adhere to. On the other hand, the way in which we could determine what the moral obligations are of firms in day-to-day situations is turned into a much more contextual enterprise that needs to take into account the particular efficiency gains of the situation in question. Heath notes that:

> *...actually making the case [for specific moral obligations] requires a more detailed analysis, one that examines the specific conditions of the market in question. These remarks are clearly unsatisfactory. The more general research program, however, is one that I believe has considerable promise.* (Heath 2014, 41)

6.3 A Metaethical Objection to Deriving Moral Obligations from the Model of the Perfectly Competitive Market

Although Heath tries to separate the way in which we can determine day-to-day moral obligations from the conditions specified by the model of the perfectly competitive market by means of his bottom-up approach, the justification of these obligations can still be found in said model. It is at this point in the argument that my objection comes in.

The model of the perfectly competitive market is an ideal whose conditions can never be fully met. Firms can thus never completely follow the requirements set out by Heath's ethic. This is something Heath recognizes and tries to work around. For example, he says that firms need to "minimize negative externalities" because "without some pollution there would be no economy" (Heath 2014, 36: 37). These market failures unavoidably occur because the set of property rights is not complete (as not everything in the world is owned or "can be owned," such as most parts of the sky and the sea) and because all actors on the market do not have access to all necessary information (Heath 2014, 35). As we can readily see, a minimalization of pollution is not a complete elimination of negative externalities. Although we wouldn't be able to reach such an elimination in reality, such an elimination of negative externalities is demanded of us by the model of the perfectly competitive market. This means that firms are put in an impossible situation. On the one hand, they are supposed to (in the end) eliminate their negative externalities, but on the other hand, it is impossible to avoid such market failures in practice.

Heath tries to work around this problem by saying that companies should therefore only try to minimize negative externalities, but a basis for such a proviso cannot be found in the model of the perfectly competitive market.[6] Its Pareto-optimal conditions demand the end of these negative externalities, among other things.[7] The same holds true for Friedman's business ethics, as it tries to ground obligations in the model of the perfectly competitive market along similar lines. As long as the model of the perfectly competitive market is held to be the sole foundation for the ethical obligations of firms (and no "general morality" or any other purposes are brought in), these ethical obligations are ultimately void.

In this regard, the model of the perfectly competitive market would demand all firms to become the idealized entities that the model theorizes. As economists readily recognize, these idealizations were never meant to become a possible reality. However, by setting the model of the perfectly competitive market as the foundation of ethical obligations, these impossibilities are still, in the end, demanded of firms.

If the government would grant firms their unique privileges under the market failures approach, then they would be put in a morally impossible position. Because firms aren't the idealized entities of the world of perfect competition, the state would be asking them to strive for an impossible ideal. Such a demand goes contrary to the very nature of morality, because it would not allow firms to ultimately choose between moral and immoral actions. This negates the very fact that makes morality possible, which is our volition (Binswanger 1981, 8). To be moral, an action needs to be voluntarily chosen from a set of alternatives. Because the world can never be fully consistent with the model of the perfectly competitive market, firms are, in the end, not allowed such an alternative. This means that as an ethic, the market failures approach cannot get off the ground because it fundamentally derives moral obligations from an impossible standard.

Let me try to illustrate this objection with an example that was mentioned previously. Consider a person who aims to draw a mathematically perfect circle with a pair of compasses on a whiteboard. Such a person will most definitely fail in this task, because the circle that he ends up drawing will always differ from a mathematically perfect circle. This might have several causes. For example, his compasses might have shifted a little bit, or his marker might not have excreted an even amount of ink while being dragged across the whiteboard. In the end, one can always argue that the atoms that make up the circle are vibrating a bit because of heat energy. This means that in reality, one is always unable to draw a mathematically perfect circle. The implication of this fact is that it is impossible to derive moral obligations from the task of drawing a mathematically perfect circle. Someone tasked with such a goal would never be able to achieve it, as it cannot be successfully achieved. Consequently, such a task is outside the realm of morality. The attempt to derive

[6] Heath says that society needs to accept this minimal pollution in exchange for the goods that are produced, but no argument is given that would explain how this consideration would be connected to the grounding of moral obligations in the model of the perfectly competitive market Heath (2014, 36). As it stands, it seems that this proviso runs counter to Heath's idea that moral obligations are solely derived from the model of the perfectly competitive market and do not involve general morality.

[7] For more examples, see Heath (2014, 37).

moral obligations from the model of the perfectly competitive market proceeds analogously, as it asks people to commit to an abstraction that is impossible to realize in reality.[8]

When discussing possible excusing conditions for immoral behaviour by firms, Heath seems to be aware of the kind of objection that I am posing here:

> *...it cannot be argued that these demands [of the market failures approach] are too onerous in principle, since the demands simply articulate the way that capitalist economies are supposed to function in the first place.* (Heath 2014, 38)

Heath holds that firms are supposed to function according to the model of the perfectly competitive market because that is what the government should demand of firms. However, this does not take into regard the point that the demands set upon firms by said model are in principle *impossible*.[9]

It seems that the only way that Heath could work around this objection is by bringing in other moral principles or purposes that would allow him to mitigate the moral impossibilities that are caused by the model of the perfectly competitive market. The downside to such a defence seems to be that if he would attempt it, then his approach would collapse into the kind of stockholder theory that (partly) derives moral obligations from general morality. This is problematic for Heath's approach because he sees it as an important innovation of his approach that it does not need to appeal to general morality (Heath 2014, 90).

In order to concretize this possible workaround, let us relate it to the example of the person that tries to draw a mathematically perfect circle. We now ask him to draw a mathematically perfect circle that is good enough. But the question that then arises is: *Good enough with respect to what?* Any attempt to limit the precision of a mathematically perfect circle needs to be justified by some kind of outside consideration. For example, one might say that the circle needs to be good enough for people to see that it is a circle instead of an oval. In such a case, one has brought in a purpose that comes from outside the model of the mathematically perfect circle.[10]

---

[8] Note that the difference between the task of trying to draw a mathematically perfect circle and Heath's market failures approach is just the way in which one determines practical action. As discussed in Heath's critique of Friedman's approach, Heath argues that his ideal cannot be achieved linearly (whereas the approximation of a mathematically perfect circle can be achieved linearly). This is beside the point of my example, however, as it serves to illustrate the nature of the abstraction that is being used instead of the nature of the way it can be approximated. For an analysis of why the type of abstraction that the model of the perfectly competitive market utilizes is faulty, see Reisman (1998, 425–437).

[9] As I will argue at the end of this section, the fact that firms can fundamentally never behave in line with the model of the perfectly competitive market is a failure of the descriptive power of the model. It seems to me not a surprise that firms cannot conform to an inaccurate description of their behaviour, but an inversion of the relationship between theory and reality.

[10] Bear in mind that the standard for perfection in this case has shifted because its purpose has shifted from a purely mathematical to a visual purpose. A regular circle drawn on a whiteboard with a pair of compasses is visually perfect, as it allows one to distinguish it from other kinds of shapes (such as ovals). Thus, the drawn circle perfectly fulfils its standard. As will be argued in the next paragraph, however, such an application is not possible when trying to realize the model of

The line of defence that is open to Heath is analogous to this example, as it would need to bring in moral principles or purposes that come from outside the model of the perfectly competitive market. But as was noted, such principles or purposes would then require their own justification and collapse the market failures approach into the kind of stockholder theory that Heath wishes to distinguish himself from.[11]

In a deeper sense, however, it is not possible to successfully execute such a defence.[12] Because market failures always inevitably occur (as Heath recognizes), human beings can fundamentally never behave as specified by the perfectly competitive market model. From an epistemological point of view, this means that the model of the perfectly competitive market does not accurately capture the nature of human volition with respect to the actual ways in which humans *can* and *do* act. This is where the analogy between the model of the perfectly competitive market and the model of the mathematically perfect circle comes apart. Because humans possess fundamentally different properties than mechanistic entities such as circles, they cannot be modelled with the same kind of mathematical idealizations.[13] The fact that market failures inevitably occur is an epistemological problem for the way in which the model tries to describe actual human behaviour, instead of an opening that can provide a basis for moral obligations. In this sense, my metaethical critique of the perfectly competitive market model hinges on a metaphysical fact (volition) that backfires on the very epistemological basis of the model once said fact is recognized.[14] The implication of this criticism is that we need a fundamentally different

---

the perfectly competitive market (which partly shows why the model is faulty). See Binswanger (1981) for an extended defence of this conception of perfection.

[11]As discussed in footnote 6, Heath seems to say that the inevitable market failures that 'slip through the cracks' of government regulations and moral obligations need to be accepted by society in exchange for the goods produced. But why should society accept this exchange? It seems that any attempt to answer this question would require a defence along the lines that I have suggested, as it would involve moral principles or purposes that come from outside the model of the perfectly competitive market.

[12]I thank Péter Róna for pressing me on this issue.

[13]Let us note again that one can draw a perfect circle, if one recognizes that the concept of perfection then becomes a function of visual aptness (i.e., being flawless when seen with the naked eye) instead of a function of mathematically infinite precision. However, one cannot shift the function of the adjective 'perfection' in the model of the perfectly competitive market, as said model is supposed to be a direct standard for actual markets Heath (2014, 39–40). The implication of this point is that the type of abstraction that the model of the perfectly competitive market engages in differs fundamentally from that of the mathematically perfect circle. For an analysis and critique of the type of abstraction that the model of the perfectly competitive market engages in, see Reisman (1998, 425–437).

[14]Note that that the model still has value as a game-theoretic scenario that is a logically sound way of describing the activity of abstract actors. However, this activity cannot be properly called 'competition' and claiming that it is only leads to the confusion that the model serves as a standard for what competition both is and should look like. This means that a renaming/rebranding of the model of the perfectly competitive market is necessary. I thank Brendan Hogan for gently pressing me on this issue.

way of looking at competition that can successfully describe how people *can* and *do* behave. Here, it is important to note that the model of the perfectly competitive market is not the only game in town.[15]

6.4 Conclusion

These considerations lead to the conclusion that the model of the perfectly competitive market is not a suitable source for moral obligations and, by extension, for a moral defence of markets. If we wish to preserve the metaethical principle that moral actions need to be voluntarily chosen from a set of alternatives (and, as a corollary, take into account volition), then we need an approach for looking at and modelling competition that does not abstract away from essential aspects of human nature. In order to judge and defend the operations of the market, we need epistemological and ethical theorizing that can correctly conceptualize human action and does not result in a type of idealized abstraction that cannot be realized in reality.[16]

References

Binswanger, Harry. 1981. The Possible Dream. *The Objectivist Forum* 2 (1 & 2).
Friedman, Milton. 1962. *Capitalism and freedom*. Chicago: University of Chicago Press.
Heath, Joseph. 2014. *Morality, competition, and the firm: The market failures approach to business ethics*. New York: Oxford University Press.
Lipsey, Richard G., and Kelvin Lancaster. 1956. The general theory of second best. *Review of Economic Studies* 24 (1): 11–32.
Reisman, George. 1998. *Capitalism: A treatise on economics*. Danvers: Jameson Books.
Simpson, Brian P. 2005. *Markets don't fail!* Lanham: Lexington Books.
———. 2010. Two theories of monopoly and competition: Implications and applications. *Journal of Applied Business and Economics* 11 (2): 139–151.

[15] These remarks are clearly unsatisfactory as a full critique of the model of the perfectly competitive market, but I hope to have shown the fundamental point at which the model is faulty as a result of the fact of volition. See Simpson (2010) for an alternative conception of competition (competition as rivalry) and a broader critique of the model of the perfectly competitive market that shows other facts and considerations that the model does not accurately take into account.

[16] This paper has benefitted from comments on previous drafts by Tara Smith, Péter Róna and Manon Abbo, and from the comments of the attendees of my presentation of a previous draft at the conference 'Economics as a Moral Science,' organized by the Polish Economic Institute. I would also like to thank Hanno Sauer for giving the course that led me to write this paper.

Chapter 7
Research Ethics in Economics: What If Economists and Their Subjects Are Not Rational?

Altug Yalcintas and Eylül Seren Kösel

Abstract Economists and their subjects are not always rational. The problem is so significant that economics is not able to satisfy the principal criterion for science that several philosophers of science have formulated since the logical positivists in the Vienna Circle in the 1920s: In order for a theorem to be meaningful, it has to be confirmed by the facts of the world. The increasing frequency of unverified and invalidated theories in economics suggests that economics suffers from the consequences of a questionable research practice in the processes of scientific knowledge production – the practice of refusing to reject theories that are invalidated by hard evidence and counter argumentation.

7.1 Introduction

> In response to the observation that the shocks are imaginary, a standard defense invokes Milton Friedman's (1953) methodological assertion from unnamed authority that "the more significant the theory, the more unrealistic the assumptions." More recently, "all models are false" seems to have become the universal hand-wave for dismissing any fact that does not conform to a favorite model. The noncommittal relationship with the truth revealed by these methodological evasions and the "less than totally convinced ..." dismissal of fact goes so far beyond post-modern irony that it deserves its own label. I suggest "post-real."
> Romer (2016a)

In a well-known article, Thomas Herndon et al. (2013a) replicated two of the papers by Reinhart and Rogoff (2010a, b). In their papers, Reinhart and Rogoff argued that "median growth rates for countries with public debt over roughly 90% of GDP are about 1% lower than otherwise; average (mean) growth rates are several percent lower." Herndon, Ash, and Pollin found that the works of Reinhart and Rogoff featured coding errors (especially where no data were available for some of the European countries), exclusion of data (of Australia, Austria, Belgium, Canada,

A. Yalcintas (✉) · E. S. Kösel
Economics Department (Mülkiye), Ankara University, Ankara, Turkey
e-mail: altug.yalcintas@politics.ankara.edu.tr

P. Róna et al. (eds.), *Words, Objects and Events in Economics*, Virtues and Economics 6, https://doi.org/10.1007/978-3-030-52673-3_7

and Denmark), and unconventional summary statistics. They claimed that "selective exclusion of available data, coding errors and inappropriate weighting of summary statistics lead to serious miscalculations that inaccurately represent the relationship between public debt and GDP growth among 20 advanced economies … when properly calculated, the average real GDP growth rate for countries carrying a public-debt-to-GDP ratio of over 90% is actually 2.2%, not −0.1% as published in Reinhart and Rogoff. That is, contrary to [Reinhart and Rogoff], average GDP growth at public debt/GDP ratios over 90% is not dramatically different than when debt/GDP ratios are lower." Austerity policies, they concluded, were unnecessary. Papers by Reinhart and Rogoff have been two of the papers frequently referenced by those who argued for the austerity plans in Europe and the US after the 2008 Financial Crisis. In other words, the proposal(s) for the necessity of austerity plans relied on a paper full of errors and miscalculations. Reinhart and Rogoff (2013) responded to their critics in a newspaper article and accepted that their papers involved errors. However, the papers were not retracted from the *American Economic Review* and the *National Bureau of Economic Research Working Papers*. The editors of the *American Economic Review* rejected their work.[1] Herndon et al. (2013b) published their paper in the *Cambridge Journal of Economics* in 2013.

7.2 Primum Non Nocere

Scientists can cause harm in many ways.[2] They plagiarize.[3] They also fabricate and falsify data (Fanelli 2009). If you think these are all an economist needs to know about harm in science, you might be seriously wrong (Necker 2014). Working papers published on the websites of Replication Network[4] and Replication Wiki[5] have shown that the authors of a significant number of articles in economics have not disclosed their data and computer codes for independent researchers to test the model of the original article. Papers presented at conferences[6] and special issues of

[1] Personal correspondence with Thomas Herndon (September 2018). See also Jakob Kapeller's editorial in the Heterodox Economics Newsletter, 3 March 2014.

[2] Plagiarism, (data) falsification, and (data) fabrication (or manufacturing) have considered to be the most common and significant forms of research misconducted by the authorities of research integrity around the globe, including the Office of Research Integrity (US), the National Institutes of Health (UK), and ALLEA – All European Universities (EU).

[3] See the RePEc Plagiarism Page available online at https://plagiarism.repec.org [Accessed March 2019]. See also Enders and Hoover (2004) and Karabag and Berggren (2012).

[4] Replication Network: https://replicationnetwork.com; [Accessed March 2019].

[5] Replication Wiki: http://replication.uni-goettingen.de/wiki/index.php/Main_Page [Accessed March 2019].

[6] See for instance "the International Workshop on Scientific Misconduct and Research Ethics in Economics" (organized by Association for Social Economics and Ankara University, held in Izmir, Turkey, August 2014), "Replication in Microeconomics," "Meta-analysis of Reproducibility in Economics Research," and "Replication and Ethics in Economics: Thirty Years after Dewald,

economics journals[7] report that a significant portion of the previously published works that are replicable are either unverifiable or invalidated. Blogposts published by Retraction Watch[8] suggest that economists have continued to cite the articles that were retracted by the journals in which the articles were published.

The number of academic articles and books on research ethics in economics are continuously growing. Since the 1990s, economic methodologists such as Deirdre N. McCloskey, Arjo Klamer, Stephen T. Ziliak, James R. Wible, and George F. DeMartino, among others, have expressed concerns regarding a persistent problem in applied economics. They argued that the economic profession has ignored "the ethical challenges that attend the profession's influence over the lives of others."[9] Economic methodologists have also claimed that the issues related to the scientific integrity in economics have become a pressing issue since the 2008 Financial Crisis.[10] The foundations of economic science are now seriously challenged by the cumulative consequences of a general absence of accountability and responsibility in economic research. A growing interest among economists in research integrity in economics suggests that research ethics will soon be the next turn in economic methodology where the consequences of questionable research practices will be discussed and analysed.[11]

Our primary goal in this paper is to introduce and develop this new focus. We think that the theories that cannot be verified and validated by empirical data should be abandoned (Yalcintas 2016). Economists who do not abandon unverified and invalidated theories give rise to an ethical problem in the profession. Economists who do not abandon the theories when theories are unverified or invalidated by empirical facts do not behave according to the "standards for intellectual honesty," as Imre Lakatos (1970) once put it. As the case of Reinhart and Rogoff shows, however, economists stick to their guns when they are challenged by counterevidence and refuting argumentation.

We claim that economics should be a part of a system of research ethics.[12] Research ethics is a field of study in which scholars examine the harmful consequences of researchers who are involved in questionable research practices. As

---

Thursby, and Anderson" (organised by the American Economic Association, held in Chicago, IL, January 2017), and many other events organized by the Association for Integrity and Responsible Leadership in Economics and Associated Disciplines. See the AIRLEAP website at http://www.airleap.org [Accessed October 2018].

[7] See, for instance, the *Review of Social Economy*, Special Issue on "Scientific Misconduct and Research Ethics in Economics" (2016). See also the calls for papers by *Energy Economics* and the *Journal of Economic Psychology* that will publish special issues on replication studies in 2018 or later.

[8] Retraction Watch: https://retractionwatch.com/category/by-subject/economics/ [Accessed October 2018].

[9] DeMartino and McCloskey (2015).

[10] DeMartino (2011a, b), Dow (2013).

[11] Wible (2016), Yalcintas and Wible (2016).

[12] See Kapp (1963, 3–4) who was among the first economists who put forth this argument.

David B. Resnik (2015) argues, "[m]any different disciplines, institutions, and professions have norms for behavior that suit their particular aims and goals. These norms also help members of the discipline to coordinate their actions or activities and to establish the public's trust of the discipline." In the absence of scholarly norms, self-interested researchers do not always cause the epistemic welfare of other scholars to grow. Scholarly norms include responsibility, accountability, respect, trustworthiness, and other scientific virtues that lead to a sustainable process of knowledge production. A sustainable process in science amounts to correcting errors in the first place. However, self-correction is not naturally embedded in scholarly practices. In order for errors to be corrected, the scholarly community not only need to be able to criticize the works of others. Among other things, the scholars should also be able to abandon the theories that are invalidated by counter evidence and refuting argumentation.

In the medical sciences, where there are a number of professional institutes doing research on questionable research practices, researchers are thoroughly trained before they are involved in new projects. Prestigious medical journals publish papers on research ethics as well. Economists have been inspired by biomedical scientists in many ways. However, research on questionable research practices in economics has not been common. Since the eighteenth century, economists have imitated the ways in which physicists conducted scientific research. Economists have also copied and reproduced various types of research practices in biomedical sciences. Since Dr. Quesnay who published his *Tableau Économique* in 1758, a book on which William Harvey's invention of blood circulation played a biggest role, such economists as William Petty, John Locke, Joseph Clément Juglar, amongst many others, have either been trained in or heavily influenced by medical sciences.[13] Today, biomedicine is in the process of becoming a dominant paradigm in economics. Economists use metaphors that they abduct from medical sciences, such as economic crises, toxic assets, recovery programs, healthy economies, and economic prescriptions written by IMF and World Bank. More importantly, economists' ways of conducting research resembles the ways in which biomedical scientists conduct theirs. Economists publish fewer and fewer books than articles in which an increasing number of them are being multi-authored rather than single authored. The number of citations to the work of an economist is as important as the argument of the work. As it has become more visible to the general public since the 2008 Financial Crisis, amounts of research monies granted to economists and salaries paid to economists sitting in the advisory boards of big companies of the banking industry all around the world without being accountable to the public are more or less identical to those of the pharmacologists, cancer researchers etc. In other words, types of questionable research practices in economics are similar to the types in biomedical sciences but there is no powerful sign of concern amongst economists toward internalizing the spill-over effects of questionable research practices. We claim that *primum non nocere* ("first, do no harm") is not only one of the principles that medical

[13] Groenewegen (2001), Clément (2003).

students are taught at health institutions. It should also be a condition for economists to reach their ideals. We see morality as a must-condition for economics based on facts.

Therefore, we ask: what if economists and their subjects are not rational? While we were formulating the research question, we were inspired by one of the works of Arjo Klamer, entitled "As If Economists and Their Subject were Rational" (1987), in which he argued that "[t]he real goal is not to know how we ... can design rational criteria of truthfulness; it is instead to comprehend how economists actually argue and how their argumentation works." He claims that economics is an art of persuasion and the assumption of rationality is a metaphor where economists study their subject of research (i.e. humans) as if they are rational. Most of us would argue that humans are *not* always rational. Klamer would agree. "Talk based on the pretense that everyone is rational," Klamer says, "may simply not be that interesting." However, one of consequences of not designing rational criteria of truthfulness is the loss of fact-checking and verification. In his article, Klamer does not confront the results of a fact-free world. It is obvious that a post-fact economics is not what Klamer has intended when he argued against rational criteria of truthfulness. Then, the research question becomes the following: What are the consequences of the behaviour of the economists who are not rational?

We think that the consequences of the fact that economists and their subjects are not always rational have been understudied.[14] The problem is so significant that economics is not able to satisfy the principal criterion for science and scientific theories that several philosophers of science have formulated since the logical positivists in the Vienna Circle in the 1920s: In order for a theorem to be meaningful, it has to be confirmed by the facts of the world. Today, logical positivism (or "verificationism," as many philosophers of science would like to call it) is "perceived almost universally as a villain," Wade Hands (2001,72) argues, "a wrong move that is responsible for much of what is wrong in nearly every intellectual discipline." As a result, only a limited number of economic methodologists[15] have studied the nature and consequences of the absence of verification, an ideology that Karl Popper (1935/2002, 18) once thought is "logically inadmissible." We disagree. We argue that the increasing frequency of unverified and invalidated theories in economics suggests that economics suffers from the consequences of a questionable research practice in the processes of scientific knowledge production – the practice of refusing to reject theories that are invalidated by hard evidence and counter argumentation.

[14] Necker (2016), Hoover (2006).

[15] Ward (1972), Hutchison (1984), Machlup (1955 and 1984).

7.3 Post-factual Economics

It has been more than a decade since the 2008 Financial Crisis hit the global economy. There is now a growing economic literature on the nature and consequences of the crisis. One of the widely accepted views about the crisis has been that it was not only a crisis of the global economy; it was a crisis of the economics profession as well.[16] Economists have played roles in the emergence of the crisis.

However, we have found that research ethics has not been commonly taught in economics departments around the globe since then. The number of papers published in the journal of economics education and economic methodology is also very limited.[17] In our view, it is not unfair to say that the issue of immorality among economists has been disguised in the abstract debates in economic methodology since the 2008 Financial Crisis.[18] We think that being suspicious about hard facts and evidence has played a role in excluding ethics from methodological debates in economics. As Boumans and Davis (2010, 171) argue "[t]he most important methodological value judgments in economics involve three kinds of choices made by economists regarding how economics should be carried out: (i) the choice of the subject matter to be investigated; (ii) the method to be used in investigating that subject matter; and (iii) the criteria standards, and norms used to assess and judge the validity of the investigation's outcomes." Indeed, theory rejecting is as important as theory choosing. The former is an ethical matter whereas the latter is methodological.

One of the first sources that helped the economists to focus their attention on the nature of the 2008 Financial Crisis was *Inside Job* (2010), a Hollywood documentary, directed by Charles Ferguson. This documentary not only explained how the complex financial tools worked. It also gave an overview of the period of deregulation from the 1980s to date. The view that the documentary held was that big companies such as Goldman Sachs, Morgan Stanley, and Lehman Brothers knew that the crisis was coming; but they did not do anything about it. The documentary was shown in Cannes, Toronto, and New York film festivals in 2010. It also won the 2010 Academy Award.

Inside Job is the first-ever Hollywood production that touches upon the issue of research ethics in economics. Ferguson argues that the economics profession is fully corrupted. The documentary provides evidence that economists write reports to financial companies and consulting agencies, but they do not always express it openly that they do. According to Ferguson, this leads to the issue of conflict of interest in economics.

Conflict of interest in the academe is a big issue.[19] But it is not the only form of questionable research practices. Mathematization, amongst others, has also been an

[16] Colander et al. (2009), Kirman (2010), Elliot (2010), Lawson (2009), Harvey (2012).

[17] Yalcintas and Selçuk (2016).

[18] Maziarz (2017).

[19] Valdés (1995), Mirowski (2002), Perkins (2004), Easterly (2006).

issue that economic methodologists have focused their attention on in their criticisms. For instance, Jesus M. Zaratiegui (1999) argues that "[m]athematics is a powerful symbol of the internal logical consistency that economics has developed during this century. Nevertheless, it has been accused of making a non-critical use of mathematical methods and of converting these methods into a weapon of economic imperialism." Likewise, in two of his most recent papers, the 2018 Nobel Memorial laureate Paul Romer argues that macroeconomic theory suffers from the lack of empirical evidence to support the mathematical models of economic growth. He calls it "mathiness." Mathiness, according to Romer, is a misuse of mathematics where economists do not provide the reader with the facts of the world to back up the theoretical model. Romer (2016b) thinks that non-existent empirical content in economics "signals a shift from science to academic politics" causing a scientific failure:

> The style that I am calling mathiness lets academic politics masquerade as science. Like mathematical theory, mathiness uses a mixture of words and symbols, but instead of making tight links, it leaves ample room for slippage between statements in natural versus formal language and between statements with theoretical as opposed to empirical content.

Romer is not against the use of unrealistic assumptions in economics. He is concerned, however, with the fact that mathematical macroeconomic models cannot be verified because the models are not tested whether they fit to the true facts of the world. Romer's views on macroeconomic theory, according to our understanding, are a reformulation of the general discontent among many economic methodologists who have pointed out the risks of the overuse and misuse of mathematical techniques in applied economics since the WWII. Paul Romer, a macroeconomist who has been known for his mathematical contributions to the literature on endogenous growth theory, made the point although he did not provide evidence in some of his most cited papers, such as "Endogenous Technical Change"[20] and "Economic Integration and Economic Growth",[21] to support his mathematical models, either. But he is as clear as any professional economic methodologist:

> My conjecture is that string theory and post-real macroeconomics illustrate a general failure mode of a scientific field that relies on mathematical theory … conformity to the facts is no longer needed as a coordinating device. As a result, if facts disconfirm the officially sanctioned theoretical vision, they are subordinated. Eventually, evidence stops being relevant (Romer 2016b).

[20] Romer (1990).

[21] Rivera-Batiz and Romer (1991).

7.4 The "New" Dichotomy

Economics profession has long suffered from fact-free theorizing. The arguments "proven" with some tests of statistical significance[22] under certain (and unlikely) circumstances have been accepted and widely used as if the models were verified by facts and data. As Toby Young (2016) put it "we are all post-truthers and probably always have been."

"We are living in a world of post-truth," says Ralph Keyes in his book, *The Post Truth- Era: Dishonesty and Deception in Contemporary Life* (2004) "Post-truthfullness exists in an ethical twilight zone. It allows us to dissemble without considering ourselves dishonest." The prevalence of factually inaccurate conclusions in economics have also drawn the attention of many economists since the 2008 Financial Crisis. The problem that we think has led to post-factual theorizing in economics is that economists do not behave responsibly while they conduct research. The main source of the lack of responsibility is that

- economists do not refer to data and facts
- they do not abandon theories when theories are refuted by data and facts

Originally, an understanding of ethics in science refers to the normativity debate on the dichotomy between "is-statements" and "ought-to-statements." The following statement is an is-statement: "The average rate of the US annual growth was over 4% between 1983 and 1988." The statement is replicable and it is produced by the researchers who are allegedly accountable at the World Bank.[23] However, the following statement is an ought-to-statement where the proposition is prescriptive and originating from an opinion: "In order for the global economy to achieve sustainable growth, it ought to introduce the free market policies of the US." The normativity debate has been popular in economic methodology since David Hume first formulated it in his *A Treatise on Human Nature* (1739). Many economic methodologists have often referred to it when they talked about "value judgments" and "morality."[24]

Here, we claim that in order for economists to remedy the problems of the age of post-factual economics, economists should reconsider the old dichotomy with new lenses. The "new" dichotomy refers to the logical distinction between a correct statement (or "is-statement") and an incorrect statement (or "is-not-statement"). Economists should insist on working with is-statements (such as "According to IMF, the global economy grew at 3.5% in 2017"[25]) whereas they should reject is-not statements (such as "… median growth rates for countries with public debt over

[22] Ziliak and McCloskey (1996, 2004, 2008).

[23] Source: https://data.worldbank.org/indicator/NY.GDP.MKTP.KD.ZG?locations=US [Accessed March 2019].

[24] For instance Blaug (1980, 49–134).

[25] Source: https://www.imf.org/external/datamapper/NGDP_RPCH@WEO/OEMDC/ADVEC/WEOWORLD Accessed March 2019.

roughly 90% of GDP are about 1% lower than otherwise; average (mean) growth rates are several percent lower"[26]). Failure to distinguish between an is-statement and an is-not-statement, which is in fact a logical condition, gives rise to ethical problems in scientific research. An is-statement is a statement where the model or explanation is evidenced by data and verified by the facts of the world. An is-not-statement, on the other hand, is a statement where the model cannot by supported by data and facts. Today, an understanding of ethics amounts to bringing data and facts back into science so that we can distinguish between correct and incorrect statements.

When economists do not reject a refuted theory, they do not behave responsibly to the audience because economists who keep using an is-not-statement inflict harm on other researchers. The condition for economists to account for the world in meaningful ways, they are required to stick to is-statements and reject is-not-statements. In order for a theory to produce a meaningful explanation, the theory should be verified with facts in the first place. Only theories that are verified by facts can lead to meaningful explanations. In other words, economics is not only value-laden, it is also fact-laden.

Fact-ladennes of a theory is a condition for the process of knowledge production to operate without interruptions. Interruption can occur when economists are not able to correct inconsistencies and manipulations. If a theory is free from facts, it is unlikely that a criticism is able to eliminate the abstract nature of the theory where the theory is not able to account for the causes of an event. Under such conditions, academic conversations can be easily locked in to ideological nonsense where the explanation does not have any relevance to the facts of the world. In other words, fact-free theories do not always produce pragmatic results about the world in which we live.

Theories that are free from facts cause opinions to replace evidence in scientific inquiry. An economist who only refers to opinions can build fictive models in which critics might be unable to spot erroneous contents. Fictive models are often able to account the facts of the world. Although the main problem is not necessarily fictive models *per se*, as the case of Reinhart and Rogoff suggests; is-not-statements that are invalidated by evidence inflict harm on the community of researchers as well as the general public.

7.5 Looking Forward

Economists have long been interested in the state of economics.[27] Back in the 1970s, Wassily Leontief focused on the misuse of mathematics in economics. In his "Theoretical Assumptions and Nonobserved Facts" (1971), he claimed that "[i]n the presentation of a new model, attention nowadays is usually centered on a step-by-

[26] The statement belongs to (Reinhart and Rogoff 2010b) and it is refuted by Herndon et al. (2013b).

[27] Babbage (1830), Marx (1859), Veblen (1909), Hoover (1926).

step derivation of its formal properties … By the time it comes to interpretation of the substantive conclusions, the assumptions on which the model has been based are easily forgotten … What is really needed, in most cases, is a very difficult and seldom very neat assessment and verification of these assumptions in terms of observed facts." In fact, the finding of Reinhart and Rogoff that "median growth rates for countries with public debt over roughly 90% of GDP are about 1% lower than otherwise; average (mean) growth rates are several percent lower" was, in Leontief's terms, a "nonobserved fact." Therefore, one of the questions that economic methodologists often asks is the following: What is wrong with economics?[28] Economists, to a large extent, believe that ethics is irrelevant to what they are doing in the labs, classrooms, and conference halls. Especially applied fields of economics in which economists use data and algorithms to account for what they think is important for the economy is where morality and ethics are least requested.

As The Economist (2016) put it straightforwardly, "humans do not naturally seek truth. In fact, as plenty of research shows, they tend to avoid it." Historically, we think that abstention from data and facts has started with David Ricardo. Ricardo was one of the first steps to also move away from the moral philosophies of Adam Smith and David Hume. When Ricardo thought that he refuted Smith's theory of value, he in fact refuted the morality that Smith's theory involved.

Abstract methods, many applied economists would tell, do not require ethics and morality. For instance, in *The Elements of Pure Economics* (1874), one of the most influential works in the history of economics, Léon Walras (1874, 39) claimed thus: "we need not concern ourselves with the morality or immorality of any desire which a useful thing answers or serves to satisfy." Milton Friedman (1953, 180–213) agreed with Walras when he argued that: "Positive economics is in principle independent of any particular ethical or normative judgements." Economics requires excellence in using the tools that an economist is equipped with. Besides all, there are certain mechanisms that deal with the consequences of immoral behaviour. If you are a doctoral student, you advisor will monitor your actions. But if you are professional economist, does your department or your faculty hold you responsible for your actions at the universities at all times? The evidence suggests that this is always the case. We think that the unresponsiveness among economists has a lot to do with the belief that "truth will out." Indeed, why do economists need to concern themselves with "the morality or immorality" of a subject if the truth, as Hercule Poirot (David Suchet) said it in one of the episodes of *Agatha Christie's Poirot* (2013), "has the habit of revealing itself"? It is this belief that economic methodologists should question.

[28] Fullbrook (2004), Coy (2009), Bell and Kristol (1981), Krugman (2009).

References

Babbage, C. 1830. *Reflections on the decline of science in England, and on some of its causes*. London: B. Fellowes etc.

Bell, D., and I. Kristol. 1981. *The crisis in economic theory*. New York: Basic Books.

Blaug, M. 1980 [1987]. *The methodology of economics: How economists explain*. Cambridge: Cambridge University Press.

Boumans, M., and J.B. Davis. 2010. *Economic methodology: Understanding economics as a science*. New York: Palgrave.

Clément, A. 2003. The influence of medicine on political economy in the seventeenth century. *History of Economics Review* 38 (Summer): 1–22.

Colander, D., M. Goldberg, A. Haas, K. Juselius, A. Kirman, T. Lux, and B. Sloth. 2009. The financial crisis and the systematic failure of academic economics. *Critical Review: A Journal of Politics and Society* 21 (2–3): 249–267.

Coy, P. 2009. What good are economists anyway? *Bloomberg Businessweek*, April 16.

DeMartino, G.F. 2011a. *The economist's oath: On the need for and content of professional economics ethics*. Oxford: Oxford University Press.

———. 2011b. On the need for professional economic ethics. *The Economist*, January 6.

DeMartino, G.F., and D. McCloskey. 2015. *The Oxford handbook of professional economic ethics*. Oxford: Oxford University Press.

Dow, S.C. 2013. Codes of ethics for economists: A pluralist view. *Economic Thought: History, Philosophy, and Methodology* 2 (1): 20–29.

Easterly, W. 2006. *The white man's burden: Why the west's efforts to aid the rest have done so much ill and so little good*. Oxford: Oxford University Press.

Elliot, L. 2010. Rescuing economics from its own crisis. *The Guardian*, November 8.

Enders, W., and G.A. Hoover. 2004. Whose line is it? Plagiarism in economics. *Journal of Economic Literature* 42 (2): 487–493.

Fanelli, D. 2009. How many scientists fabricate and falsify research? A systematic review and meta-analysis of survey data. *PLoS One* 4 (5): 1–11.

Friedman, M. 1953 [1984]. The methodology of positive economics. In *The philosophy of economics: An anthology*, ed. D.M. Hausman, 180–213. Cambridge/New York/Sydney: Cambridge University Press.

Fullbrook, E. 2004. *A guide to what is wrong with economics*. London: Anthem Press.

Groenewegen, P.D. 2001. *Physicians and political economy: Six studies of the work of doctor-economists*. London/New York: Routledge.

Hands, D.W. 2001. *Reflection without rules: Economic methodology and contemporary science theory*. Cambridge: Cambridge University Press.

Harvey, J.T. 2012. How economists contributed to the financial crisis. *Forbes*, June 2.

Herndon, T., Ash, M., & Pollin, R. 2013a. *Does high public debt consistently stifle economic growth? A critique of Reinhart and Rogoff*. PERI working paper series: 322.

Herndon, T., M. Ash, and R. Pollin. 2013b. Does high public debt consistently stifle economic growth? A critique of Reinhart and Rogoff. *Cambridge Journal of Economics*. https://doi.org/10.1093/cje/bet075.

Hoover, G.E. 1926. The present state of economic science. *Social Forces* 5 (1): 57–60.

Hoover, G.A. 2006. A game-theoretic model of plagiarism. *Atlantic Economic Journal* 34 (4): 449–454.

Hutchison, T.W. 1984. On verification in economics. In *The philosophy of economics: An anthology*, ed. D.M. Hausman. Cambridge/New York/Sydney: Cambridge University Press.

Kapp, K.W. 1949 [1963]. Economics as part of a system of applied ethics. In *History of economic thought: A book of readings*, ed. K.W. Kapp. New York: Barnes and Noble.

Karabag, S.F., and C. Berggren. 2012. Retraction, dishonesty and plagiarism: Analysis of a crucial issue for academic publishing, and the inadequate responses from leading journals in

economics and management disciplines. *Journal of Applied Economics and Business Research* 2 (3): 172–183.
Keyes, R. 2004. *The post-truth era: Dishonesty and deception in contemporary life*. New York: St. Martin's Press.
Kirman, A. 2010. The economic crisis is a crisis for economic theory. *CESifo Economic Studies* 56 (4): 498–535.
Klamer, A. 1987. As if economists and their subject were rational. In *The rhetoric of the human sciences: Language and argument in scholarship and public affairs*, ed. J.S. Nelson, A. Megill, and D.N. McCloskey, 163–183. Madison: University of Wisconsin Press.
Krugman, P. 2009. How did economists get it so wrong? *The New York Times,* September 6.
Lakatos, Imre. 1970. Falsification and methodology of scientific research programmes. In *Criticism and growth of knowledge*, ed. Imre Lakatos and Alan Musgrave, 170–196. Cambridge: Cambridge University Press.
Lawson, T. 2009. The current economic crisis: Its nature and the course of academic economics. *Cambridge Journal of Economics* 33 (4): 759–777.
Leontief, W. 1971. Theoretical assumptions and nonobserved facts. *American Economic Review* 61: 1–7.
Machlup, F. 1955. The problem of verification in economics. *Southern Economic Journal* 22 (1): 1–21.
———. 1984. On indirect verification [Rejoiner to a reluctant ultra empiricist]. In *The philosophy of economics: An anthology*, ed. D.M. Hausman. Cambridge/New York/Sydney: Cambridge University Press.
Marx, K. 1859. *Zur Kritik der Politischen Ökonomie*. Berlin: F. Duncker.
Maziarz, M. 2017. The Reinhart-Rogoff controversy as an instance of the 'emerging contrary result' phenomenon. *Journal of Economic Methodology* 24 (3): 1–13.
Mirowski, P. 2002. *Machine dreams: Economics becomes a cyborg science*. Cambridge: Cambridge University Press.
Necker, S. 2014. Scientific misbehavior in economics. *Research Policy* 43 (10): 1747–1759.
———. 2016. Why do scientists cheat? Insights from behavioral economics. *Review of Social Economy* 74 (1): 98–108.
Perkins, J. 2004. *Confessions of an economic hitman*. Ebury Press.
Popper, K.R. 1935 [2002]. *The logic of scientific discovery*. London/New York: Routledge.
Reinhart, C. M., Rogoff, K. 2010a. Growth in a time of debt. NBER discussion paper no. 15639.
Reinhart, C.M., and K. Rogoff. 2010b. Growth in a time of debt. *American Economic Review: Papers and Proceedings* 100: 573–578.
Reinhart, C.M., and Rogoff, K. 2013. Debt, growth, and austerity debate. *New York Times*, April 25.
Resnik, D. B. 2015. *What is ethics in research, and why is it imporant?* National Institute of Environmental Health Sciences. https://www.niehs.nih.gov/research/resources/bioethics/whatis/. Accessed Mar 2019.
Rivera-Batiz, L.A., and P. Romer. 1991. Economic integration and endogenous growth. *Quarterly Journal of Economics* 106 (2): 531–555.
Romer, P. 1990. Endogenous technical change. *Journal of Political Economy* 98 (5): S71–S102.
———. 2016a. The trouble with macroeconomics. *The American Economist*. https://ccl.yale.edu/sites/default/files/files/The%20Trouble%20with%20Macroeconomics.pdf. Accessed Mar 2019.
———. 2016b. Mathiness in the theory of economic growth. *American Economic Review* 105 (5): 89–93.
The Economist. 2016. Yes, I'd lie to you. September 10.
Valdés, J.G. 1995. *Pinochet's economists: The Chicago School of Economics in Chile*. Cambridge: Cambridge University Press.
Veblen, T. 1909. The limitations of marginal utility. *Journal of Political Economy* 17 (9): 620–636.
Walras, L. 1874 [2010]. *Elements of pure economics*. London.
Ward, B. 1972. *What's wrong with economics*. New York: Basic Books.

Wible, J.R. 2016. Scientific misconduct and the responsible conduct of research in science and economics. *Review of Social Economy* 74 (1): 7–32. https://doi.org/10.1080/00346764.2016.1135598.

Yalcintas, A. 2016. *Intellectual path dependence: Why economists do not reject refuted theories*. London: Routledge.

Yalcintas, A., and I.Ş. Selçuk. 2016. Research ethics education in economics. *Review of Social Economy* 74 (1): 53–74.

Yalcintas, A., and J.R. Wible. 2016. Scientific misconduct and research ethics in economics: An introduction. *Review of Social Economy* 74 (1): 1–6.

Young, T. 2016. The truth about 'post-truth politics'. *The Spectator*, July 16.

Zaratiegui, M. Jesus. 1999. The imperialism of economics over ethics. *Journal of Markets & Morality* 2 (2): 208–219.

Ziliak, S.T., and D.N. McCloskey. 1996. Standard error of regression. *Journal of Economic Literature* 34 (1): 97–114.

———. 2004. Size matters: The standard error of regression in the American Economic Review. *Journal of Socio-economics* 33 (5): 527–546.

———. 2008. *The cult of statistical significance: How the standard error costs us jobs, justice, and lives*. Ann Arbor: University of Michigan Press.

Chapter 8
Economic Choice Revisited: Lessons from Pre-modern Thinkers

Agnieszka Wincewicz-Price

Abstract This paper explains why economics needs a better understanding of human moral agency. It argues that more thorough conceptualisation of economic choice as human choice exposes what is missing in the neoclassical account of the economic agent. The view of human agency presented here is derived from Aristotle, Thomas Aquinas and Adam Smith. It is a view rooted in the virtue ethics tradition, one that highlights the concept and value of practical reason for agency.

8.1 Introduction

The prevailing view of standard economic theory is that economists need not engage in the analysis of moral agency. For purposes of explanation and prediction, we are told, neoclassical economic theory requires that behaviour of the economic agent under investigation should merely conform to a set of axioms whose plausibility is mostly theoretical. To the limited extent that the behaviour of persons corresponds with these assumptions, economic activity can be explained in the framework of rational choice.

In what follows I argue that broader and therefore better understanding of human choice can help improve the quality of economic explanation. While the uniqueness of a human economic agent conceived of as a moral person surpasses the limitations of the text-book economic theory of choice, it nonetheless plays an important role in informing some of its premises.

The view of human agency I want to offer for the consideration of economists is one that is rooted in the Aristotelian-Thomistic tradition, important parts of which, as I will try to show, are shared by Adam Smith. It is a tradition marked by virtue ethics and development of practical wisdom. This tradition's anthropological claims about human action expose what is missing in the neoclassical account of the economic agent.

A. Wincewicz-Price (✉)
Polish Economic Institute, Warsaw, Poland

P. Róna et al. (eds.), *Words, Objects and Events in Economics*, Virtues and Economics 6, https://doi.org/10.1007/978-3-030-52673-3_8

The way I approach the moral aspects of human choice is intended to engage critically some core elements of modern economic theory of preferences. The word 'moral' in this context refers to human rational agency which encompasses one's ability to reason, to formulate one's goals, values and principles of action, and to pursue one's goals in line with those principles or values.

The structure of the paper is as follows. The next section discusses some problematic elements in the standard economic understanding of human agency and choice. Particular attention is given to the lack of nuance in the treatment of ends and inexistence of a sound theory of ends formation in this theory. The following section encapsulates the essence of what human moral agency meant in the pre-modern virtue-ethics tradition. Directions for possible application of this broader approach to human choice in economics are suggested next. Final section concludes.

8.2 Neoclassical Account of Economic Agency

8.2.1 Economic Agent – 'Not Human But Important'[1]

From its very inception as a subject of enquiry, economics has been studied as the product of human agency under conditions of scarcity. These were understood to be real actions of real human persons. With its emancipation from the theological and philosophical vision of man, modern economics gradually lost interest in the moral aspects of human action and agency. Already in the seventeenth century, economics and moral philosophy separated into two distinct paths of enquiry. Nonetheless, the human person as a moral agent continued to be treated in one way or another as an important factor of economic analysis. The famous definition of economics proposed by Alfred Marshall in the late nineteenth century retained the spirit of that approach. In his *Principles of Economics* he called it 'a study of mankind in the ordinary business of life' which 'examines that part of individual and social action which is most closely connected with the attainment and with the use of material requisites of well-being' (Marshall 1890, 1).

Since around the middle of the twentieth century, however, there has been a tendency in economics (as well as in other social sciences), to avoid explanation of phenomena, which originate in human action, by appealing to the conscious decisions of persons in favour of impersonal factors (Evans 2002, 14). As the subject of economic science and analysis, human beings have increasingly been losing their distinctiveness as persons, as moral agents, being replaced by impersonal economic agents, necessarily non-moral. It is the latter, not the former that is considered the core subject of today's economic analysis.

[1] Cf. Ross (2012).

In the course of modernisation and specialisation, the discipline of economics emancipated itself from a broader discourse of moral science. As a result many important normative concepts associated with human choice and decision making were left outside of economic analysis. With the arrival of the modern world, modern economics was born whose assumptions of human conduct originated arguably in the ideas about human rationality and reason which were developed in the Enlightenment period (Langholm 1992, 1–10).

The concept of agency employed in standard micro and macroeconomics is one of the rational agent. Rational agency in the economic sense originates directly from the axioms of rational choice theory and it should not be confused with other concepts of rationality employed especially in philosophy or ethics. The neoclassical concept of the economic agent derives from the work of Samuelson (1947). A Samuelsonian economic agent, as Ross (2005, 245) defines him, is simply 'any system that observes certain consistency conditions in its behaviour, such that it can be interpreted as if it is maximizing the value of a function that maps a system of preferences over commodity bundles onto the real numbers'.

Standard rationality theory, or expected utility theory, is axiomatic in nature, in that it is formulated around a specific set of logical assumptions (completeness, transitivity, independence, and continuity) chosen in order to produce well-defined preference orderings (von Neumann and Morgenstern 1944). The 'von Neumann-Morgenstern utility function theorem' states that 'if these four axioms are satisfied, any set of well-ordered preferences can be represented by a distinct (monotonic) individual utility function'. The utility function is the theory's representation of the agent. The individual is defined as no more than a collection of preferences (Davis 2006, 70).

Ross argues that nothing in economic theory requires that economic agency be identified with individual persons, because economists use it to build abstract models of firms, nations, labour unions, consortia in auctions, lineages in evolutionary games, and other feedback-sensitive, incentive-driven systems that have no evident psychological properties at all. Economic agents are thus 'representative' optimizers, whose ontological status is indeterminate. Davis (2006, 70), for example, notes that there is nothing in the current understanding of choice that requires it to apply to any type of agent in particular:

> The representation of choice now dominant in economics, which involves nothing more than maximisation of a formal objective function, applies equally well to any type of agent, whether it be a single individual, a collection of individuals, an animal, a collection of animals, or even a programmed machine.

It is therefore not important for rational choice analysis that the economic agent is a human being, let alone a moral person. The issue of what constitutes the individuality of this agent never arises. It is not a problem that economists find interesting for their purposes. The economic agent is an abstraction, a useful but merely theoretical construct of the smallest unit of activity under analysis.[2]

Economists of the formal neoclassical school grant that they abstract from the over-determined conditions of social reality, by disregarding all but the economic motives of actors in constructing theoretical models. They do not consider it necessary that their models should correspond with concrete human natures or even a general extra-economic idea of 'human nature'. In short, neoclassical economics, with its focus on abstract rational choice theory, no longer needs a distinctive notion of a person to explain economic phenomena.

8.2.2 The Peculiar Concept of Preferences

Although nothing in the theory requires that the agents in question be persons conceived of as moral agents, this fact alone does not mean that economics is uninterested in individual behaviour. Its interest in individual behaviour, however, is usually limited. Most economists study human choice because it is a building block in the analysis of social phenomena, rather than because they are concerned with explaining individual persons qua individual behaviour. In this they rely on the abstract conception referred to as homo economicus which co-defines rationality and individuality, not in connection with human behaviour, but by reference to the logical-mathematical properties of equilibrium analysis. Thus, rational choice theory is not treated as psychological theory. The theory considers only the formal or mathematical properties of the agent's preferences. What the latter term should be taken to mean is a formal ordering relation, constructed so as to allow equilibrium analysis. Hausman defines preferences as 'total subjective comparative evaluations'. He illustrates this with the example: "To say that Jill prefers x to y is to say that when Jill has thought about *everything* she takes to bear on how much she values x and y, Jill ranks x above y" (2012, 34, italics mine). So: a preference is comparative (x is ranked above y); the comparison is in terms of value; the valuation is subjective ('how much she values …'); and it takes account of the totality of factors that the individual thinks relevant to the comparison ('everything she takes to bear on …').

As total or overall evaluations, preferences are already informed by reflection on what there is reason to do. Nature of reasons for action is not inquired into. The content and rationality of preferences themselves is considered irrelevant to the theory. What matters is that preferences are known and stable over time. In order to

[2]As Becker (1993, 402) writes: 'while the economic approach to behaviour builds on a theory of individual choice, it is not mainly concerned with individuals. It uses theory at the micro level as a powerful tool to derive implications at the group or macro level. Rational individual choice is combined with assumptions about technologies and other determinants of opportunities, equilibrium in market and non-market situations, and laws, norms, and traditions to obtain results concerning the behaviour of groups. It is mainly because the theory derives implications at the macro level that it is of interest to policymakers'.

guarantee some set of rational choices, preferences must satisfy three other conditions: they must be complete, transitive and independent of irrelevant. These conditions guarantee some 'most preferred' subset of the available options which define a rational choice. Included in the 'most preferred' subset are those options which bring about the highest level of utility – another concept of purely analytical value which has no psychological meaning.

Preferences are outcomes of comparative assessments rather than inputs into deliberation. The existence of an all-things-considered preference ordering presupposes that the economic agent has been able to compare the available options from different points of view or according to different reasons or motivations that she may have had, in order to choose among those options. Models that explain and predict choices in terms of constraints, beliefs, and total comparative evaluations do not deny that human agency and motivation are complex matters. They instead locate those complexities in theories of preference formation and change and not in theories of choice. Only this way of defining preferences makes it possible to take them as determinants of choice. The separation of deliberation about ends and choice of means, however, is mostly of analytical use and has little to do with how most human decisions are made.

Such separation can only apply to deliberations wherein objects of preferences can be unequivocally compared. And therein lies the rub. In such deliberations the agent has some single, well-defined goal or function (a dominant end); or the differing goals which he pursues have some common factor, such as the satisfaction of a known desire. This approach has no use in the—not uncommon—situations in which one needs to combine two or more different goals, virtues or standards, which he feels he cannot ignore or downplay but which seem to demand incompatible things of him. In the standard economic approach, a person experiencing different motivations, who may find it impossible to compare all options, and also be unable to make a clear-cut decision between them, would be considered a case in which the agent's preferences are not fully formed (incomplete). And so he or she is incapable of a rational action. An example of such a case is a choice one has to make between integrity or charity, on the one hand, and some other incompatible goal one might have, say, the pursuit of a suitable retirement, or subsistence, or approval of others (Taylor 1985, 236–7). One can compare the strength of one's desire to have a cup of coffee now with one's desire to have a cup of tea now, and the degree of respective enjoyments and satisfactions. But how can either of those desires and their satisfaction be compared with one's desire to be a fine scholar, a good father, a true friend? Such alternatives are incommensurable, much like vacationing at the beach vs. vacationing in the mountains (Finnis 2011, 424), vs. not vacationing at all and having more money in the bank. Each possibility has some intelligible appeal not found in what makes the other appealing. It is often impossible coherently to combine all the demands that we might consider important and valid. An agent can see one reason as outweighing another only if the two reasons are contained within a single evaluative perspective. While admirable in its generality, however, this theory has almost no descriptive content. More complex incidents of human choice which involve incommensurate options cannot be easily explained in the strict terms of rational choice theory.

Thus, economists in the neoclassical tradition begin their analysis of choice at the point where ends are (or are assumed to be) set. In the Humean spirit they conceive of rationality in an instrumental manner: their task is to specify structures that are appropriate means of achieving those ends, whatever they may be. They do not account for ends which emerge from deliberations about incommensurate desires or evolve from the processes of one's maturation and learning from experience. The main reason for this is the neoclassical economics' agnosticism about values. The formation and ranking of human ends is intimately connected to values. Desires can only make sense in relation to one's idea of the good. Otherwise we cannot explain what we find valuable in them, and deciding between two or more conflicting desires is even more problematic. Choice, empirically speaking, involves not merely instrumental but also practical reason. When we try to decide what to do, or when we explain our own conduct or conduct of others, we invoke reasons for action. Yet, the assumption of stable and invariable fundamental preferences makes the discussion about values entirely redundant for economists (in line with the old maxim: de gustibus non disputandum est) and formation of values ceases to be economists' problem. In the rational choice framework, all values are converted into a single currency of meaningless utility.

It seems paradoxical that economic choice should not be concerned with the way real people decide, choose, and act. Lacking in its account is any meaningful definition of preferences as ends and of preference formation. This indeed is not a satisfactory achievement for a discipline which claims competence in explaining not merely economic phenomena but non-economic issues as well. Some evolutionary economists go as far as to claim that selection and ordering of ends proves to be the most fundamental economic problem and lament the abandonment of the problem by orthodox economics.[3]

Economists explain their ignorance in the realm of human ends with the claim that this is subject for other social scholars' authority such as anthropologists, sociologists, and philosophers. In what follows I suggest that better grasp of human ends should not be contracted to other disciplines but revisited in the spirit that was dominant in pre-modern economic thought on which its modern successor was built. I propose that the pre-modern thinkers, who did not have a separate notion of an economic agent, but who analysed all human decisions and actions in broader ethical terms, can offer important insights into what is involved in economic choice qua human choice. They show that the former cannot be completely and cleanly separated from the latter without some meaningful loss. Just how much is missed in the simplified account of economic choice can be illustrated by a short sketch of a

[3] See for example Fudulu (2014a, b). Fudulu claims that even Ludwig von Mises (1949 ,92–93) and Lionel Robbins (1932, 12) produced merely tautological definitions of ends, while Frank Knight who famously stated that ends are 'the most obstinate unknown of all unknowns' (2009, 12) questioned the value of making any inquiries into this obscura realm.

much more thorough understanding of reason and choice, one that is based on an old and well-established conception of practical reason.

8.3 Pre-modern Economic Thought: Self-Development and Practical Reason

Contrary to the analytical framework of modern economics, it was the focus on the person that located political economy among the moral sciences in the tradition of Western political philosophy. What are now called the human sciences (economics, sociology, etc.) were traditionally referred to as the moral sciences. The mark of this tradition, from Plato and Aristotle, throughout Christian antiquity, the Middle Ages (especially following Aquinas), and the Scottish Enlightenment of Adam Smith, is its placing of economic thinking within the area of moral actions of human beings. It is this tradition which informed an alternative conception of reason to that which prevailed in the Enlightenment. As Hollis (1977,12; italics mine) notes, 'there was another voice in the Enlightenment, a voice that argued that the power of reason allows humans to master nature, manipulate society, change culture and, indeed, *shape our own selves*'. The capacity for shaping our own selves is a distinctly human feature which signifies our capacity for deliberative self-determination and indicates a dynamic and undetermined character of human goals. It was given a special focus in classical philosophy and especially in the virtue ethics tradition as the below cursorily discussed works of its important representatives will show.

Virtue ethics is the oldest and broadest stream of ethical theory in the study of human conduct. Introduced by Plato and Aristotle, it was taken up by the Stoics, used by Cicero, and finally adopted into Christianity in its most lasting form by Thomas Aquinas. In this tradition, right action is defined as that which leads to the 'well-being' of the individual. The term 'eudaimonia' is a classical Greek word, commonly translated as 'happiness', but perhaps better described as 'well-being' or 'human flourishing' or 'good life'. The essence of eudaimonia consists in one's striving toward excellence based on one's unique potential. At the centre of the eudaimonistic ethics is the view of the person as a being conscious of oneself and one's own will, concerned not only with the question of what one should do but also of what one should and could desire. In the language of modern economics we would say that the person is concerned with what her ends should be, what she should value. Thus, in the Aristotelian-Thomistic tradition, moral behaviour is not only about fulfilling one's duty or obligation to others and following rules. It is just as much a theory of the formation of the will, and a development of desires and aspirations (Koslowski 2006). One does not need to accept Aristotle's theory of the will or even regard the will itself as a meaningful concept (there is a powerful modern trend to treat it as obsolete) to recognize some truth in this observation.

Aristotle, Thomas Aquinas, and Adam Smith are three particularly important representatives of this tradition. What connects the three philosophers is an explicit account of human nature characterized by man's striving for the good and self-formation ('man' was then used, but especially in Aquinas's case, 'human being' is what was meant). Their insights offer a potentially valuable addition to what has been only partially recognized by economic utility theory, decision theory, and other rationality studies. They provide building blocks for a richer concept of the economic agent conceived of as a moral person whose preferences are not fixed and stable but forming and changing. The concept of changing and undetermined preferences highlights the dynamic aspect of human agency and suggests that it is not merely satisfaction of preferences but the very process of their formation which makes human agency different from non-human agents and constitutes part of people's welfare. Contrary to the instrumental means-ends approach of the neoclassical economic theory of choice, this process calls for the exercise of practical reason by the person in question – a distinct feature of human agency.

8.3.1 Practical Wisdom in Aristotle

The great influence of Aristotle not only on ethics, but also on what later came to be called economic thought,[4] makes his works a natural starting point of enquiry into the moral nature of an economic agent and economic choice. Aristotle is the first thinker to analyse economic life systematically, so that it is seen as 'embedded' in the extended fabric of the community. For Aristotle, economic affairs do not have a free-standing status but properly belong to practical philosophy. Unlike modern economics, he treats of economic phenomena such as production, consumption, exchange, in the context of virtues. Aristotle's anthropology stresses the importance of man's capacity and propensity to develop and practise virtue, and accounts for a person's responsibility for acquiring a good character. His 'economic analysis', which is usually identified in Book V of the Nicomachean Ethics and Book I of the Politics, is only a sub-section within an enquiry into other, more essential subject-matters (Finley 1970, 5). The most important criterion of evaluation of these various matters appears to be the opportunity for each agent to realize his own potential and thus to fulfil himself in a flourishing life. When reading Aristotle's reflections on what we call today economic or market behaviour, one cannot resist the impression that economics for him is mostly about human values, concerned more generally about eudaimonia (Pol, I, 9, 1257b, 40-1258a, 2).

What are today referred to as economic pursuits, though important, are for Aristotle simply one among other means toward human flourishing and excellence. One's endeavours related to securing some level of material status have a functional, not finalistic, nature and it is a necessary, but not sufficient, instrument for attaining a good life. It is true that a good life cannot do without the possession of material

[4]The Scholastics used the Nicomachean Ethics as one of the leading textbooks. Some important contributions of such influential modern thinkers as Marx and Keynes were derived from Aristotle's works.

goods. Nonetheless, it is not exhausted by the material component, but depends, rather, on a plurality of human dimensions. Present in Aristotle but missing in modern perception of the economic agent is the motivational complexity of every human choice and thus of economic choice, too.

Aristotle is famous for his strongly teleological perspective according to which the highest human good involves activities that are goal-directed and have purpose. Most importantly, the essential end point (telos) is to achieve the best that is within us. As paraphrased by Johnston (1997, 6): 'The excellence of the human being is thus going to be associated with growth towards some final realization of his or her true and best nature.'

For Aristotle, an agent is roughly defined as a being who can entertain, and take steps to fulfil, 'reasonable desires'. For him, the rationality that is the distinguishing mark of agency is neither merely formal nor only instrumental. His is the rationality understood as prudence (phronesis). The 'phronimos' in Aristotle is the person who has practical wisdom (as compared to theoretical wisdom), full reasonableness (as opposed to perfect rationality). This includes the ability of discerning what things are good for himself and for mankind at the point of choice, rather than merely theoretically. Such a person is the norm of action, both economic and non-economic (NEII.6:1107a1; VI.11:1143b15: VI.5:1140b8-10).

It is because of complexity and multidimensionality of choice that Aristotle, contrary to the standard economic approach appreciates that deliberation about man's ends does not benefit from the means-end paradigm (Wiggins 1980). Instead, it requires practical wisdom. For Aristotle, it is the mark of the man of practical wisdom to be able to 'deliberate well about what is good and expedient for himself, not in some particular respect (health, strength) but about what sorts of things conduce to the good life in general' (NE, Book VI). Deliberation in Aristotle is not primarily a search for means, but rather a search for the best specification of the good that one wants. A nontechnical deliberation is one in which man has a vague description of something he wants—a good life, a satisfying career, a fulfilling hobby, etc. The problem is not to see what will be causally efficacious in bringing this about, but rather to see what really qualifies as an adequate and practically realizable specification of what would satisfy this want. As long as this specification is not available, there is no room for means. This observation is not given enough consideration in the mainstream economics account of human choice.

Aristotle's account of human choice and deliberation indicates the unfinished or indeterminate character of our ideals and value structure, which are constitutive both of human freedom and practical rationality itself. The life of virtue consists in intellectual and moral virtue, i.e. in the ability to discern what is best and the ability to act on this judgment. In other words, moral virtue serves to bring our actual choices and preferences into line with what our better judgment recommends to us (Veatch 1962, 101). Aristotle's account of deliberation indicates that the human choice situation cannot always be treated as forming a closed, complete, consistent system.

One of the purposes of wise decision making is to discover what we prefer or what is really of value. For it is only through wise and morally imaginative deliberation that we gain clarity about those objects we have previously deemed to be of value. Wise decision making does not assume that we already have a fixed and absolute set of preferences or values:

> Matters concerned with conduct and questions of what is good for us have no fixity. The agents themselves must in each case consider what is appropriate to the occasion (NE 1103b 26).

In Aristotle (and, as will be shown below, also in Aquinas) a person is conceived through his actions as *en route* toward some completion (Schenck 1976). Aristotle's good man develops over time (Burnyeat 1980, 69). Virtue is not something we are born with but something we acquire. In a way, the whole Nicomachean Ethics is Aristotle's reply to the question of how we grow up to become the fully adult and self-sufficient rational animal that is the end toward which the nature of our species tends. This process of self-completion and all that it entails reveals the human person to be a potential and not a fully actual being.

8.3.2 Aquinas and the Scholastics: The economic Realm as Space for Personal Development

Influenced by Aristotle, the scholars of medieval universities in Europe known as the Scholastics adopted a similar approach to the study of human action. The moral dimension of human behaviour served as a basis of what was later called scholastic economics.[5] Most medieval writers who ventured into the discussion of economics saw it as a form of behaviour which like all others must be considered within the discussion of human beings as of God's creation and, as such, morally or law-bound actors. Scholastic anthropology claims that persons are beings capable of moral (self-)rule, whose sense of duty is 'an important but not the only determinant of their conduct' (De-Juan and Monsalve 2006; Decock 2013; Langholm 1998).

Although economics was not yet acknowledged as an independent discipline, it formed a consistent body of doctrine according to which economic relations ought to be ruled, by the laws of distributive and commutative justice, always finally by the 'law of charity'. The economic realm is just one among other spheres of human activity where learning of virtuous conduct takes place. The anthropology underlying the scholastic paradigm in economics is not that of the homo economicus whose goal is the satisfaction of given preferences. Their famous doctrines of usury and just price are ultimately grounded in the vision of a morally responsible man whose

[5]According to Schumpeter, it was the 'Scholastic doctors' of the Middle Ages who deserve the title of founder of economics: „It is within their systems of moral theology and law that economics gained definite if not separate existence, and it is they who come nearer than does any other group to having been the 'founders' of scientific economics." (Schumpeter 1954, 93). In Schumpeter's view 'the scholastics and the natural-law philosophers had worked out all the elements of' economic analysis, while Smith simply undertook 'the task of co-ordinating them' (Schumpeter 1954, 178).

will is thought to be autonomous, capable of creating its own laws, or at least ruling itself (De-Juan and Monsalve 2006). That is one way to speak of 'free will'. But within the hierarchy of laws, the will remains subject to a plethora of heteronomous laws. Those laws include positive human laws, divine law, and natural law (Decock 2013, 523). This vision of human nature implies the possibility to act contrary to right reason and to positive law. Nothing is pre-determined or „given" but has to be reasoned through and decided by each individual.

Among these Scholastics it was particularly Thomas Aquinas who, in Schumpeter's view, contributed most to establishing the grounds of modern scientific analysis (1954, 8). Summa Theologiae sketches a broad view of Thomas's understanding of human being in which he further develops Aristotle's insights concerning moral agency. In Thomas's vision, as in Aristotle, human action is teleological or purposive, in that man's goal is to achieve the virtuous life which realizes, so far as possible, one's potential as a human being. The intellectual and moral virtues perfect the human intellect and appetite in proportion to human nature (ST Ia IIae., q. 62, art. 2).[6] In Aquinas's formulation, the human person is naturally ordered toward seeking the true and achieving the good through the operation of the intellect (reason) and the will, even as sin has corrupted this nature. Whether man seeks good or ill, there is a general orientation toward perceived goods. He does not choose what he knows to be fully bad. Whatever man desires, he desires it under the aspect of good (ST I-II, q.1).

Being human, however, means that we can be very much mistaken in our judgements and decisions. Thus, Aquinas's approach, similarly to Aristotle's, focuses on man's imperfect knowledge of the self and his fundamental uncertainty with reference to what he wants and what he should value. For Aquinas, the telos of persons is grounded in self-consciousness and self-mastery. Man always strives for increasingly more complete self-fulfilment: 'all desire the fulfilment of their perfection, and it is precisely this fulfilment in which the last end consists, as stated above' (ST I-II, q.1, art. 5). In order to become a good person, one needs to acquire the virtue of prudence, which is covered in Aquinas's notion of 'prudentia': ST II-II q.47 aa 1–7. Prudence enables us to reason well towards the choice of commitments, projects, and actions, and to apply general practical principles to concrete circumstances in order to choose rightly. Prudence involves both the choice of the end sought and the means to attain that end. It calls for reason rather than impulse; and takes counsel from others in the selection process of ends and then of means to achieve the chosen end (Elmendorf 1892, 4). In the Summa, prudentia is described as superior to theoretical knowledge, for prudence perfects the cognitive faculty.

Aquinas's conception of the person characterized by prudentia resembles Aristotle's account of the virtuous person. For Aristotle, 'the virtuous person performs the right action in the right way at the right time on the right objects' (Rorty 1980, 380), and for Aquinas it is someone 'really knowing what one is doing, being

[6] Unlike Aristotle, the ultimate goal is set beyond human nature in sharing in the divine nature which was described above as the beatific vision.

aware of the circumstances and consequences of one's actions, with the right conception of the sort of action one is performing' (Aquinas, question 61, articles 3 and 4). In both accounts the source of action is the human himself. Life of virtue is the gold standard, the end toward which man is striving.

8.3.3 Moral Maturation in Adam Smith

Adam Smith may not be the most obvious thinker to be associated with the Aristotelian-Thomistic tradition. Yet, his thought is very much in line with the concept of the human agent as a moral person developed by the two great philosophers. The sources of his moral enquiry, especially, go back to the ancient tradition of the Greeks and the Romans (Szulczewski 2015, 91). Smith's essential premise, which he shared with the ancient thinkers, was that moral science could only proceed from a thorough understanding of human nature and human agency was the root of his economic understanding (Oakley 1994, ix). His understanding of man was much broader than the one that is often mistakenly ascribed to him today. Smith believed that many behavioural regularities can be explained better by an understanding of people's attitude toward actions, rather than their valuation of final outcomes. He notes that many of our choices are 'not so much founded upon [their] utility,' but reflect primarily 'the great, the noble, and the exalted property' of the action or activity itself. Related to this, he thought that we should never investigate human nature as such, in a void, namely merely theoretically. Even if such an investigation were possible, it would prove disappointing, for it could teach us little about the origins of men's attitudes. He, in a likewise manner to Aristotle and Aquinas, puts the practical above the theoretical in most things.

Smith did not accept the static view of man as coming into the world already fully equipped to make correct moral decisions. To him, such a conception of man neglects the everyday process by which moral decisions are actually arrived at. His Theory of Moral Sentiments (TMS, henceforth), in particular, shows man's devotion to self-improvement. It accounts for the crucial process to which all people are subject, that 'of moral and psychological *development* from the early stages of childhood to that of the mature moral agent' (Brown 1994: 95; Montes 2008: 45). Indeed, Adam Smith's TMS is devoted to the question of how people form moral sentiments. The Smithian individual agent forms his moral sentiments through an informal learning process that is realized in social interactions (Thoron 2016).

Smith posits in TMS and again in WN a common human 'desire of bettering our condition' (WN II.iii.28; TMS I.iii.2.i); and implies, though he nowhere states it explicitly, a broad notion of 'betterment', including 'a moral kind that involves improving our character and has nothing to do with acquiring material goods' (Fleischacker 2004, 63; see also Raphael and Macfie 1984, 9; Heyne 2008, 59–63; Griswold 1999, 130–136; and Otteson 2002, 196–197). The desire for self-improvement, both material and moral, which Smith takes to be natural for human beings, seems incompatible with the posited behaviour of the creature called rational economic man. He saw virtue in one's self-improvement, not only in material but also, moral aspect. He says, 'upon many occasions…

> [what] prompts us to the practice of those divine virtues is not the love of our neighbour, it is not the love of mankind (…). It is a stronger love, a more powerful affection, which generally takes place upon such occasions; the love of what is honourable and noble, of the grandeur, and dignity, and superiority of our own characters [italics added].

It is not so much the calculation of potential gains, as the desire to do proper things, to do things that would meet the approval of a disinterested by-stander, the impartial spectator (Choi 1990, 293); God, even. Choi (1990, 294) posits, in a Smithian vein, that it is the awareness of the presence of others like oneself and moral judgments made with this awareness that make a person a human being. Knowledge of what constitutes the proper thing is not always obvious and often requires deliberation. In Smith this deliberation is supported by what he calls the impartial spectator: the hypothetical observer who passes positive or negative judgments upon the actions of those around him, directed by virtuous considerations, whether of the intellectual or some other kind; an imaginary construction that individuals call upon as they evaluate their own sentiments and conduct. We endeavour to examine our own conduct as we imagine any other fair and impartial spectator would examine it. The impartial spectator signifies one's ability to distance oneself from his particular situation and subjective point of view, and to see oneself from a perspective that he can share with others. It also emphasizes the desire for approbation as a key fact of the human condition.

Smith argues in TMS that it is not the love of praise and attention that motivates moral actions, but the love of praiseworthiness. Though we care about the opinions of others, we care much more about the opinions of the impartial spectator and want to avoid his disapprobation. The 'man within the breast' is thus often seen as the key to Smith's account of the faculty of conscience:

> The jurisdiction of the man without, is founded altogether in the desire of actual praise, and in the aversion to actual blame. The jurisdiction of the man within, is founded altogether in the desire of praise-worthiness, and in the aversion to blame-worthiness (TMS, III, i, 31–2).

In Storr's phrasing, the impartial spectator is 'the imaginary figure that each of us constructs to offer us moral guidance as we *negotiate* our lives' (Storr 2013, 3, emphasis added). The use of the word 'negotiate' is of great importance here. Many of our decisions need judgment. Judgment, in turn, is an outcome of a particular type of dialogue we exercise with the impartial spectator as we negotiate various aspects of our lives, or – to put differently – as we try to determine what ends we ought to pursue.

The impartial spectator is both a guide as we experience the world and is affected by our experiences in the world. The refinement of the impartial spectator marks our maturation as agents. That is why, in practical reasoning, it is not assumed that we act from a well-defined base of knowledge but rather that this knowledge is constantly in a process of formation, which in turn depends to a large extent on our experience—for instance, as to whether it is a more or less wise spectator. As an outcome of a dialogic process, such judgment is 'freely subscribed to and freely acted upon, and cannot be predetermined or rule-bound'.

Smith, like Aristotle, understands moral judgment as the exercise of 'practical wisdom necessary for right actions in particular situations' (Hanley 2009, 87). He discusses on a number of occasions how our knowledge of our own best course of action comes from our experience as opposed to solely abstract calculation. Smith's moral agent is not someone who calculates choice based solely on a set of motivations. His capacity of calculation is imperfect, and his motivations are dynamic and constantly under formation. The inclinations to choose one concrete goal over another and the activity of evaluating others' and one's own behaviours are both formative processes for Smith, not initial assumptions. This can be seen in his extensive discussions of the formation of knowledge and the formation of desires, values, and interests.

Economic theory treats the formation process as an aside or an afterthought to the judgment about action, while for Smith it is the judgment itself that forms our objectives—meaning the whole process of judging—as it is also the means we take to realize them.

Whereas economists in the neoclassical tradition analyse choice under a full-information assumption (Levy 1995, 305), Smith's theory of the impartial spectator indicates that any person's preferences, values, and purposes are neither given nor stable. They are unknown fully at any one moment, but not ultimately unknowable. The epistemic limitation is a result of man's limited self-experience. The eventual (partial) knowability is a result of careful and ongoing observation of one's own moral life as if one were a third-party spectator. A critical mark of Smith's theory is that man needs to learn to see his interests in their true perspective. Finding out what those interests are, however, is a process that economists are unable to capture if they assume given preferences. As Fleischacker (2004, 61, 63) notes, '[w]hen we ask after the 'nature' of human beings we are looking for what human beings 'really' want, beneath the surface trappings (…) Human nature always includes what people aspire to, for Smith; it is never reduced [as in the economist's version of utilitarianism] to the desires they merely happen to have.'

The view of the human agent that emerges from the above discussion of Aristotle, Thomas Aquinas, and Adam Smith is a picture of a human being directing himself in relation to his own good by practical reason (phronesis in the Greek, and the virtue of prudentia for Aquinas). When a person acts on the basis of practical wisdom it means his choices are embedded in an ongoing process of self-formation. His decisions shape his judgment of his capabilities and interests, which then inform his future decisions. He does not possess complete and accurate knowledge of himself and others, but he is constantly in a process of (self-)development. His wants and needs are not certain and given determinants of his actions, as they are in the mainstream economic assumptions. It is rather his actions which from his unique experience influence his judgment and are conducive to the formation of his future goals.

8.4 Towards a Better Explanation of Choice

Rational choice theory explains decisions and choice 'as if' they were brought about by rational economic agents. While successful in domains which do not need to take account of complexity of human choice (e.g. basic consumption decisions) this approach has significant costs for the study of less obvious decisions of the morally constituted person in the economic realm. One can go so far as to say that its limiting assumptions have deprived the human economic agent of almost everything that could make it a person. It has denied the economic agent the ability to learn about what is good for him and to make value choices. The formal strictures of this theory make it impossible for it to capture an image of an evolving self, indicative of a conscious change in one's values and preferences. In short, it has neglected his capacity of practical reason. For purposes of analytical rigor and predictive power, it has abstracted this capacity away, as if it were not there. Important aspects of human agency which are not conceivable in the rational choice framework are left outside of the picture. Failing to take account of the nature of human ends it has seriously limited its descriptive and predictive capacity.

And yet, although the above analysed factors of human choice do not meet the strictures of economic rationality, they could nonetheless serve as important objects of economic analysis. For they inform many consumer, investment, and entrepreneurial decisions, as well as choices involved in our work relations and relationships with ourselves and the environment and other people. These situations call for decisions which are not—or at least not exclusively—driven by instrumental means-end reasoning. They may be better characterized by the agent's openness to unforeseen possibilities and learning. As the above necessarily sketchy accounts of the great pre-modern thinkers show, formation of ones ends – the key category in rational choice theory – is not possible without practical reasoning. This is particularly visible in problems of preference conflict, instability and uncertainty about one's goals.

There are many examples of conflicts among preferences, and of preference change, that lead to behaviour that is very puzzling from the point of view of received economic theory. They show that richer and less rigid conceptions of agency are needed to account for the fact that a person may have not just a single set but competing sets of preferences; or even contradictory preferences or preference sets. Moreover, meta-preferences which rank the preference sets, and a person's preferences, may oscillate between alternative states. Decision making based on practical wisdom rejects the notion that all alternatives can be compared on a single measure of utility. Rather, this process entails an exploration of the multiplicity of incommensurable values embedded within the available options.

Unlike the abstract homo economicus, persons have the capacity to form, revise, and change their preferences as they see fit.[7] They can even have no preference at all! While the mainstream theory assumes complete and transitive preferences

[7] 'Choice', as John Dewey (2008, 96) put it, 'signifies a capacity for deliberately changing preferences.'

which implies that all aspects of a choice setting can be weighed off against each other and all alternatives can be ranked in terms of 'betterness', people can sometimes refuse to do this. Real-world persons often do not, cannot, or will not reach judgements about some things.

Likewise, the idea of rational economic choice is an inadequate representation of human choice in conditions of an unknown or not fully specified goal. Whereas rational choice theory paints a picture of a goal-oriented agent, guided by known and stable references, human decision making is not always based on a well-defined goal. As the above discussion shows, the moral person seeks to construct a good life for himself and those who depend on him guided by a vision of the good which he is open to revisit based on new experience. Through the exercise of practical reason, he develops and improves his vision of the good and the good life.

The process of self-formation and development of one's idea of the good should be regarded as part of human moral agency. Obvious general examples are quitting an addiction or taking up a challenge to become physically fit. Moral human agent is someone who strives for self-betterment through conscious formation and change of preferences—meaning, constant refinement of the preferences, and the reasons for (continued) preferring of something or someone over other possibilities. Without enough experience or at least reflection about what the goal entails, we cannot provide an explicit description of how to arrive at it. We often choose from many imagined futures and can rarely expect with certainty that what we choose will work out. Through open-ended involvement in the various practices of our life we learn to exercise a practical judgment, a judgment that becomes operative in practice but which we cannot always give a fully specified or systematic account of. We nonetheless act on it. It enables us to express what is valuable for us and to form ourselves in accordance with what we find valuable.[8]

What makes persons distinct economic (and moral) agents is thus not to be found merely in the content or character or solidity of preferences, but rather in their instability and process of formation. Unlike nonhuman agents, humans possess a reflective capacity which enables them to form their preferences and make judgments about their preference orderings, as well as change these preferences as a result of a change in their values and purposes. In less formal language we can say that only human choice is preceded by attempts to sort out: (1) what is worth doing (choice of ends), and (2) what sort of person to be or become (as a distinct end). This shows that moral agency is not to be defined only by its capacity for altruistic or other-regarding sentiments but also by one's striving for self-betterment through a conscious formation and change of preferences in line with chosen ends. As such, the human agent (the person) is more complicated and less certain about his preferences than the traditionally-conceived economic agent. Such agent can better be portrayed not by more elaborate interpretations of what brings utility, but by accounting for the uncertainty and instability of his preferences.

[8] Similar ideas are exercised by Charles Taylor (1985) in his notion of human identity, which is defined by our fundamental evaluations, that is articulations of what is worthy, more fulfilling, etc.; and Harry Frankfurt in his essay 'The Importance of What We Care About' (1982).

This implies, however, the need to relax the two above-mentioned economic assumptions: (1) that preferences are given and (2) that they are stable. In fact, behavioural economists have also questioned these assumptions. Many aspects of human choice, which were traditionally regarded as insignificant in standard neoclassical theory, have been gaining importance. This is challenging the prevailing view that there is no need for economists to delve into the complex issues determining human choice. It is now more readily admitted that better understanding of how people choose can help develop better ideas about what contributes to their welfare and design better policies. A more complex picture of agency has been inspiring departures from mainstream micro- and macroeconomic models.[9] These new developments, however, stop short of connecting their conclusions to the moral nature of human agency. Not only does the formation of preferences depend strongly on the many psychological and cognitive factors that do not affect choice according to the rational choice approach (the framing-effect of the problem, the predetermined default rules which prefer certain selection decisions, the way risks are conveyed etc.), but it is also affected by conflicts of interests, values, wishes and desires. Clear and obvious long-term preferences, say, good health, can come into conflict with short-term preferences, such as yielding to temptation and having another cigarette. The value of caring for an old parent may be incompatible with long hours spent at work to maintain one's family. Choices like this are not made easier by correcting for behavioural biases. They instead require practical reason and, indeed, moral wisdom.

Practically speaking, the strictures of rational choice theory also turn out to be particularly difficult for welfare policy. If policy advice is to be based on the promotion of individual well-being, the advice that comes from assuming that people are well-behaved preference satisfiers cannot apply to a world where individuals fail to reveal well-behaved and well-defined preferences. Since real persons fail to behave as predicted by expected utility theory, because their choices reveal incoherent preferences, their behaviour cannot be meaningfully represented by a utility function. Welfare economists thus lose evidence concerning that which promotes their welfare, in all but the most basic or general preference sets (water, shelter, food, clothes…). In other words, they do not have an obvious way to determine what people's 'true' preferences are, or – to be more precise – what they value (Hargreaves-Heap 2013).

Better understanding of how people choose and what constitutes their welfare requires that human agents are understood and explicitly treated as moral, and not merely economic, agents. For, as I tried to illustrate, moral considerations are inseparable aspects of economic activity. They co-determine the choice behaviour of economic agents and thereby direct the outcome of economic activity. Abstracting them away or forcing them into the strictures of rational choice theory limits the extent to which many contexts of human choice can be meaningfully explained and translated into policy.

[9] Cf. Simon (1957), De Grauwe (2012).

The above discussion also shows that moral agency cannot be treated as an extra ingredient in the existing framework of economic explanation. Moral agency as understood in this paper is an element which challenges the very fundaments of the rational choice framework. People do not only care about what happens. They are not mere consequentialists. Driven by various non-instrumental considerations, they also care how and for what reason things happen. From the perspective of a moral agent a choice is rational if the chooser can acceptably answer the question 'Why did you make that choice? or 'What is the value of your choice?' In this perspective, inability to grasp and articulate the importance of various goods – i.e. reasons – for one's conduct is unreasonable, if not plain irrational.

It appears then that part of what it means to be rational involves also deciding about or defining one's conception of the good and acting on it. Forming a rational response to one's environment, rejecting preferences one cannot satisfy, or developing preferences that are in line with one's principles, depends on one's idea of the good. It depends on one's values, which one learns and endorses or rejects over time while making choices. An important aspect of human agency is to be found in one's pursuits of underdefined goals. Persons define themselves and their good (at least partially) through the exercise of this form of agency.

Even in decisions which to an outside observer may seem as primarily driven by reasons based on means-end calculation, people are often motivated by reasons which fall outside of this framework of instrumental reasoning. Consumption which is traditionally regarded by economists as a textbook case of means-ends thinking could instead be seen as an act of discovering one's needs, one's values, and even self-formation. Similarly, many entrepreneurial decisions which involve high levels of risk and creativity are difficult to explain ex ante merely as the entrepreneur's motive of profit maximisation. If economists want to better understand how people form their preferences, outside of the most trivial contexts, they would benefit from a better understanding of the 'non-economic springs of economic action' (Hirschman 2013).

8.5 Conclusion

The moral character of human choice is inconvenient for the orthodox economic perspective. There is no place and no logic within this perspective for incomplete and changeable preferences, which in turn precludes any possibility to correctly understand choice. The neoclassical perspective on preferences remains agnostic about the reality of human values which inform their desires, wants and ultimately their choices.

One can wonder, however, whether economists really do not need criteria for distinguishing human individuals as distinct and independent beings, who do not dissolve into social aggregates and patterns. Is there really no use for economic theory to acknowledge that persons count because they are not reducible to economic agents? That their choices also count as their own precisely because the free will and self-formation cannot be reduced to instrumentally conceived economic agency?

Pre-modern virtue ethics tradition and its account of the place of economic activity in a broader context of human moral life provides some valuable insights into how various elements constitutive of the human choice could be better integrated. While rational-choice account of human agency is of purely formal, instrumental and arguably value-free character, the notion of practical reason in the pre-modern tradition of Aristotle, Aquinas and Adam Smith, takes a distinctively normative question as its starting point. It typically asks, of a set of alternatives for action none of which has yet been performed, what one ought to do, or what it would be best to do. It is thus concerned not with matters of fact and their explanation, but with matters of value, of what it would be desirable to do. In practical reasoning, agents attempt to assess and weigh their reasons for action, the considerations that speak for and against alternative courses of action that are open to them. Part of what it means to be rational involves also deciding about or defining one's conception of the good and acting on it. Insofar as these 'hidden' aspects of choice are important in the explanation of choice and the analysis of welfare, they should not be overlooked in economics.

References

Aquinas, Saint Thomas. 1270/1947. *Summa Theologicae*. Trans. The Fathers of the English Dominican Province. Benzinger Bros. Edition.

Aristotle. 1980. *Nichomachean Ethics*. Trans. D. Ross, Rev. J. L. Ackrill, and J. O. Urmson. Oxford: Oxford University Press.

———. 1988. *Politics*, ed. S. Everson. Cambridge: Cambridge University Press.

Becker, G.S. 1993. *Human capital: A theoretical and empirical analysis, with special reference to education*. New York: Columbia University Press.

Brown, V. 1994. *Adam Smith's discourse: Canonicity, commerce, and conscience*. London/New York: Routledge.

Burnyeat, M.F. 1980. Aristotle on learning to be good. In *Essays on Aristotle's ethics*, Amélie Oksenberg Rorty, 69–92. University of California Press.

Choi, Y.B. 1990. Smith's view on human nature: A problem in the interpretation of the wealth of nations and the theory of moral sentiments. *Review of Social Economy* 48 (3): 288–302.

Davis, J. 2006. The normative significance of the individual in economics. In *Ethics and the market: Insights from social economics*, ed. Betsy Jane Clary, Wilfred Dolfsma, and Deborah M. Figart, 69–83. London/New York: Routledge.

De Grauwe, P. 2012. *Lectures on Behavioural macroeconomics*. Princeton: Princeton University Press.

Decock, W. 2013. *Theologians and contract law: The moral transformation of the Ius Commune*. Martinus Nijhoff Publishers.

De-Juan, O., and F. Monsalve. 2006. Morally ruled behaviour: The neglected contribution of scholasticism. *The European Journal of the History of Economic Thought* 13 (1): 99–112.

Dewey, J. 2008. *The later works 1925–1953*. Vol. 3, 1927–1928. Carbondale: Southern Illinois University.

Elmendorf, J.J. 1892. *Elements of moral theology*. New York: James Pott & Company.

Evans, C.S. 2002. *Preserving the person: A look at the human sciences*. Vancouver: Regent College Publishing.

Finley, M.I. 1970. Aristotle and economic analysis. *Past & Present* 47: 3–25.

Finnis, J. 2011. *Natural law and natural rights*. Oxford: Oxford University Press.

Fleischacker, S. 2004. *On Adam Smith's wealth of nations: A philosophical companion*. Princeton: Princeton University Press.

Frankfurt, H. 1982. The importance of what we care about. *Synthese* 53 (2): 257–272.

Fudulu, P. 2014a. *An anti-entropic evolutionary perspective on human ends formation*. https://www.researchgate.net/publication/260835541_An_Anti-Entropic_Evolutionary_Perspective_on_Human_Ends_Formation. Accessed 4 June 2018.

———. 2014b. *The economic decoding of the phenomenon of culture based on the Antientropic formation and ranking of human ends: Definition, formation and impact*. https://www.researchgate.net/publication/260835377_The_Economic_Decoding_of_the_Phenomenon_of_Culture_Based_on_the_Antientropic_Formation_and_Ranking_of_Human_Ends_Definition_Formation_and_Impact. Accessed 4 June 2018.

Griswold, C.L. 1999. *Adam Smith and the virtues of enlightenment*. Cambridge: Cambridge University Press.

Hanley, R.P. 2009. *Adam Smith and the character of virtue*. Cambridge: Cambridge University Press.

Hargreaves-Heap, S.P. 2013. What is the meaning of behavioural economics? *Cambridge Journal of Economics* 37 (5): 985–1000.

Hausman, D.M. 2012. *Preference, value, choice, and welfare*. Cambridge: Cambridge University Press.

Heyne, P. 2008. In *'Are economists basically immoral?' And other essays on economics, ethics, and religion by Paul Heyne*, ed. G. Brennan and A. Waterman. Indianapolis: Liberty Fund.

Hirschman, A.O. 2013. *The passions and the interests: Political arguments for capitalism before its triumph*. Princeton: Princeton University Press.

Hollis, M. 1977. *Models of man: Philosophical thoughts on social action*. Cambridge: Cambridge University Press.

Johnston, I. 1997. *Lecture on Aristotle's Nicomachean ethics*. https://www.oneeyedman.net/school-archive/classes/fulltext/www.mala.bc.ca/~johnstoi/introser/aristot.htm. Accessed 20 Feb 2017.

Knight, F.H. 2009. *The ethics of competition*. New Brunswick/London: Transaction Publishers.

Koslowski, P. 2006. *The economy of happiness*. https://ssrn.com/abstract=945328 or https://doi.org/10.2139/ssrn.945328.

Langholm, O. 1992. *Economics in the medieval schools: Wealth, exchange, value, money and usury according to the Paris theological tradition, 1200–1350*. Leiden/New York/Köln: EJ Brill.

———. 1998. *The legacy of scholasticism in economic thought: Antecedents of choice and power*. Cambridge: Cambridge University Press.

Levy, D.M. 1995. The partial spectator in the wealth of nations: A robust utilitarianism. *European Journal of the History of Economic Thought* 2 (2): 299–326.

Marshall, A. 1890. *Principles of economics*. London: Macmillan.

Montes, L. 2008. Adam Smith as an eclectic stoic. *Adam Smith Review* 4: 30–56.

Oakley, A. 1994. *Classical economic man: Human agency and methodology in the political economy of Adam Smith and JS Mill*. Cheltenham: Edward Elgar Publishing.

Otteson, J.R. 2002. *Adam Smith's marketplace of life*. Cambridge: Cambridge University Press.

Raphael, D.D., and Macfie, A.L. 1984. Introduction to *The theory of moral sentiments*, by Adam Smith, 1–52.

Robbins, L. 1932. *An essay on the nature and significance of economic science*. London: Macmillan and Co.

Rorty, A.O. 1980. *Essays on Aristotle's ethics*. Berkeley/Los Angeles/London: University of California Press.

Ross, D. 2005. *Economic theory and cognitive science: Microexplanation*. Cambridge, MA/ London: MIT Press.

———. 2012. The economic agent: Not human but important. In *Philosophy of economics*, ed. U. Mäki, 691–736. Amsterdam: Elsevier.

Samuelson, P.A. 1947. *Foundations of economic analysis*. Cambridge: Harvard University Press.

Schenck, D. 1976. Re-casting the 'ethics of virtue/ethics of duty debate. *Journal of Religious Ethics* 4: 269–286.

Schumpeter, J.A. 1954. *History of economic analysis*. New York: Psychology Press.

Simon, H. 1957. A behavioural model of rational choice. In *Models of man, social and rational: Mathematical essays on rational human behavior in a social setting*. New York: Wiley.

Smith, A. 1776/1904. *An inquiry into the nature and causes of the wealth of nations*, ed. Edwin Cannan. London: Methuen & Co., Ltd

———. 1976. *Theory of moral sentiments*. The Glasgow edition of the works and Cor-respondence of Adam Smith, Vol. 1. Oxford: Oxford University Press.

Storr, V.H. 2013. The impartial spectator and the moral teachings of markets. In *The Oxford handbook of freedom*, ed. David Schmidtz and Carmen E. Pavel, 456–474. Oxford: Oxford University Press.

Szulczewski, G. 2015. Krytyka 'etyki sympatii i 'problem A. Smitha w swietle lektury Teorii uczuc moralnych. *Prakseologia* 157: 79.

Taylor, C. 1985. Self-interpreting animals. In *Human agency and language. Philosophical papers: Volume 1*, 45–76. Cambridge: Cambridge University Press.

Thoron, S. 2016. Morality beyond social preferences: Smithian sympathy, social neuroscience and the nature of social consciousness. *Psychology and Economics. Historical Perspective* 6 (2): 235–264.

Veatch, H.B. 1962. *Rational man: A modern interpretation of Aristotelian ethics*. Indianapolis: Liberty Fund.

von Mises, L. 1949. *Human action. A treatise on economics*. Chicago: Contemporary Books, Inc.

von Neumann, J., and O. Morgenstern. 1944. *Theory of games and economic behavior*. Princeton: Princeton University Press.

Wiggins, D. 1980. Deliberation and practical reason. In *Essays on Aristotle's ethics*, ed. A. Rorty. Berkeley/Los Angeles/London: University of California Press.

Chapter 9
Between Individual and Collective Rationality

Anna Horodecka and Liudmyla Vozna

Abstract The paper raises the question of irreducibility of collective rationality to individual rationality. The irreducibility of collective rationality to individual rationality is explained by the phenomenon of complexity and complex character of human nature. Taking the complexity theory approach to the analysis of institutions, it discusses the question of dependency of individual rationality on collective rationality. It is asserted that collective rationality emerges not merely from the human capacity for rational reasoning but from a variety of other human capabilities which influence the formation and functioning of socioeconomic institutions. Institutions, in turn, are understood here to be a specific embodiment of collective rationality.

9.1 Introduction

The concept of rationality is one of the fundamental and most controversial elements of economic theory. Since the rationality assumption forms the basis of many, primarily mainstream, macroeconomic models, it influences macroeconomic policy. Thus theories of rationality can have far-reaching effects on economic reality.

Generally speaking, economic rationality is associated with optimisation and efficiency. In modern economics, the concept of rationality is primarily related to the neoclassical doctrine which focuses on the behaviour of individuals (consumers, firms) and assumes that their rational strategy consists in the maximization of their expected subjective utility. This is exactly a core of the mainstream theory of rational choice, which is also associated with positivism, instrumentalism and methodological individualism. However, despite the popularity of the neoclassical approach, the concept of rationality has been approached from various angles and, in the literature, one can find various definitions and classifications of rationality. For example, Herbert Simon (1955) made an important distinction between "substantive

A. Horodecka (✉)
Department of Economics (Collegium of Economic Analysis), Warsaw School of Economics, Warsaw, Poland

L. Vozna
Independent researcher, Zhytomyr, Ukraine

P. Róna et al. (eds.), *Words, Objects and Events in Economics*, Virtues and Economics 6, https://doi.org/10.1007/978-3-030-52673-3_9

rationality" and "procedural rationality", paying special attention to the latter. Other authors distinguish between methodological and material (or between formal and practical) rationalities. In the analytical sense, the definition of rationality has two components: the first one concerns the rational choice of goals, and the second one – the means to realize the goals.[1] In addition to that, some leading economists, notably Vernon Smith (2008), distinguish constructivist and ecological forms of rationality.

The list of other possible classifications can be continued. Since the economic system as a complex system consists of many different levels (e.g., micro-level, meso-level, macro-level), it makes sense to consider a specific hierarchy of different rationalities. These differences are determined by the goal pursued by a system under consideration, the level of a system (from individual to meta level), the time and space criteria. In this hierarchy, neoclassical rationality occupies only one of the places and, at the same time, can diverge from other types of rationality. For example, imagine a community of people which in order to survive in the short-term, in accordance with neoclassical rationality, harvests a forest for immediate sale. However, in the long-term, deforestation can cause such changes in the ecosystem (such as the problem of floods), which can then endanger further existence of this community and the costs associated with these consequences far outweigh the benefits realised from the harvesting. This example demonstrates both the contradiction between the short-term and long-term rationalities, and the inconsistency that is possible between individual and collective rationalities. The choice that seems to be rational from the individual point of view can be irrational from the standpoint of collective choice, and vice versa.

Using the simplistic understanding of rationality, we risk failing to explain the more complex phenomena such as, for example, the functioning of some fundamental social institutions, which are associated with collective rationality. The goal of the article is to demonstrate the irreducibility of collective rationality to individual rationality. Also, using the complexity theory approach, we raise the question of the dependency of individual rationality on collective rationality, i.e. we assert that the former is conditioned by the latter rather than vice versa. Alan Kirman, one of the leading specialists in complexity economics, makes a similar assumption when he writes that "it is the type of organization rather than the individual behavior that is central to achieving coordination. Once in place, the organizational structure, itself, coordinates individual activities and makes them consistent" (2010, 6).

One of the key problems addressed here concerns the question of the influence of human emotions on rational decision-making. For explaining collective rationality, if we start from individual rationality, which implies decision-making is based on a conscious (calculated) choice, we inevitably lose the emotional component of human behaviour. However, some human emotions (such as affection, confidence, curiosity, despair, fear, pride, empathy, trust, and others) play an important role in the formation of social values and the functioning of a number of social institutions.

[1] See, for example, Hogan and Marcelle (2017).

Institutions, in turn, are understood here to be a specific embodiment of collective rationality. Therefore, it is assumed that collective rationality inevitably contains a component of human emotions, and the irreducibility of collective rationality to individual rationality is also explained by the constitutive factors of complexity determined by human nature.

9.2 The Problem of Irreducibility of Collective Rationality to Individual Rationality

One of the basic premises of rational choice theory is that aggregate social behaviour results from the behaviour of individual actors, each of whom makes their individual decisions. A central concept in this theory is the principle of transitivity and the aggregation of individual preferences as a mode to define a collective, social (rational) choice. The principle of transitivity is fundamental in the ordinal utility theory and it orders the preferences of an economic agent on an ordinal scale. It means that if a consumer (economic agent) deciding between any three goods (outcomes, choice options) X, Y, and Z, prefers X to Y and Y to Z, he must prefer X to Z. This principle is considered to be an important feature of the rational behaviour of economic agent as the transitivity principle relates to the laws of logic.

Such a view of rationality is also strictly connected with the understanding of economics as a positivist science, i.e. free from subjective judgements and values. Researchers point out that early neoclassical economists "still had traces of the old honorable concern of the classical writers, like Adam Smith, for the well-being of society" (Walsh 2007), and it was Lionel Robbins who, in the 1930s, played an important role in the transformation of neoclassical economics into the science of instrumental rationality that is "value free" and related with a choice of (scarce) means for the reaching (alternative) purposes (ends) (Cedrini and Novarese 2014). Vivian Walsh (2007, 64), in particular, notes that this methodological shift occurred not without the influence of logical positivism, still popular at the time, and "the positivist claim that there was a sharp dichotomy between matters of fact (the domain of science) and values".

Yet, by following the transitivity principle to determine collective choice we risk coming to paradoxical conclusions. In fact, the Arrow's impossibility theorem, which considers voting systems, and the prisoner's dilemma demonstrate the very problem of conversion of individual preferences into desirable community-wide acts of choice. As Amartya Sen remarks, internal consistency of choice "is essentially confused, and there is no way of determining whether a choice function is consistent or not without referring to something external to choice behavior (such as objectives, values, or norms)" (2002, 122).

Also, in line with methodological individualism, choices made on the basis of individual rationality lead to the best possible allocation of resources, so they are the guarantors of systemic rationality understood as macroeconomic rationality. In other words, market outcomes, such as equilibrium or allocative efficiency are results of individual behaviour of agents who are rational maximisers of their utility (profits). In this vein, macroeconomic rationality is associated with the Pareto optimum state, under which the situation of one of the market participants cannot be improved without worsening the situation of others.

However, this assertion is not supported by empirical data and, in particular, by experimental economics. For example, on the basis of laboratory experiments with the use of computer simulations, Shyam Sunder (2002) concluded that "weak form of individual rationality, far short of maximization, when combined with appropriate market institutions, can be sufficient for the market outcomes to approach the predictions of the first fundamental theorem" (according to this theorem, under certain idealized conditions, any competitive equilibrium leads to a Pareto efficient allocation of resources), and that "markets can exhibit elements of rationality absent in economic agents". (Rationality of markets relates here in particular to the efficiency of markets, namely their ability to allocate the limited amounts of resources in a way that maximizes the satisfaction of consumers.)

Similarly, Alan Kirman challenges the approach, according to which "if we start with well-behaved individuals we will obtain well behaved aggregates" and "well behaved individuals have nicely structured behavior derived from their optimizing behavior" (Kirman 2010, 20). Using the example of the fish markets, he demonstrates that the behavioural "regularity" is more apparent at the aggregate than at the individual level. In general, in his opinion, the relationship between the behaviour of the individual participants and the market as a whole is mediated by the way in which the market is organized, i.e. the way in which the market allocates resources depends on the type of market institution (Kirman 2010, 60–66).

9.3 The Bounded Rationality Versus the Variety of Human Nature

To explain the above-introduced problem of irreducibility, it is important to at least consider the following three questions. First, individuals are not rational in the neoclassical sense. Second, even if individuals do not behave rationally in the neoclassical sense, they can reach goals that are rational. Third, the rational behaviour of individuals can lead to irrational outcomes.

9.3.1 Individuals Are Not Rational in the Neoclassical Sense

One of the features of basic neoclassical models, such as the consumer behaviour models that use the indifference curves and budget lines (elaborated by F. Edgeworth, E. Slutsky, J. Hicks), is the implicit assumption about a consumer who has perfect information. In other words, a consumer, maximizing utility and choosing a combination of the two goods (for example the cups of coffee and the pieces of cake), knows perfectly in advance the taste of these goods, his future pleasure from their consumption and so on. It seems plausible that, if a consumer, performs the same actions regularly, is devoid of the spirit of experimentation or (and) the ability to cognize the new, it is easier for him to comprehend his expected cumulative utility. However, if our consumer decides to sacrifice additional cups of coffee in the name of an additional unit of a product, which is absolutely new for him, he takes a very serious risk of disappointment. Thus, such models are actually static.

The assumption of perfect information is also implicitly linked to the assumption of the availability of time, sufficient to make the best decision. For example, in the case considered here, the consumer needs to have enough time to collect (obtain) the necessary information for making the optimal decision. Such a period of time can extend also into the past, forming sufficient experience for making decisions in the future. On the one hand, if the consumer makes a decision very quickly and does not have the necessary information at the same time, i.e. irrationally, he risks making a choice which can be illustrated as a point remote from the point that corresponds to the maximum utility (equilibrium point). On the other hand, if our consumer has an infinitely long time to exercise his best choice, he runs the risk of repeating the fate of the Buridan's ass: failing to choose between two identical piles of hay, the poor animal eventually dies of hunger. Therefore, from the point of view of the dynamics and viability of the economic agent, it should be better for him to consume less utility but be at the right time than in an effort to maximize the utility lose it altogether.

The best-known criticism of the neoclassical concept of rationality came from behavioural economists especially from Herbert Simon whose new concept of bounded rationality refers to the limited human ability to process information (resulting, among others, from the lack of time, attention and ability to concentrate). In particular, Herbert Simon critically refers to the understanding of self-interest as the most important goal and to the way of understanding rationality as a choice of preferred alternatives of action by means of a system of values enabling assessment of the results of activities.[2] His criticism refers to the realism of the assumption about full knowledge of a person about possible alternatives, as well as the physical possibility of his/her mind to process this information and the willingness to make such an assessment. For this reason, individual decision-making behaviour does not follow from a calculation of all variants and the selection of the optimal, but from the available values and criteria, which are considered by the subject as the basis for selection. The lack of realism of neoclassical assumptions manifests itself also in the fact that, for example, for a neoclassical rational man it would be rational to violate social rules (if it does not involve costs), but people often refrain from doing so. So, according to Herbert Simon, rationality of decision-making in the neoclassical sense is not possible. In place of neoclassical rationality, he proposes the concept

[2] See, for example, Simon (1997).

of bounded rationality, which accounts for a rational choice that takes into account the cognitive limitations of both knowledge and cognitive capacity. Simon (1997) emphasizes the important role of habit, which allows for the economical use of spiritual and mental efforts.

The view that "the economy is not just governed by rational actors" and "much economic activity is governed by *animal spirits*" (i.e. people have noneconomic motives), was already expressed by John Maynard Keynes in his *The General Theory* (1936). Referring to Keynes, George Akerlof and Robert Shiller, in their book *Animal Spirits* (Akerlof and Shiller 2009) have provided a detailed illustration of this idea. In particular, they demonstrate that, in making significant investment decisions, economic actors often don't behave according to prescriptions of standard economic theory. The latter, in turn, asserts that, for making rational decisions, people consider all the options available to them, consider the outcomes of all these options and how advantageous each outcome would be, consider the probabilities of each of these options, and then they make a decision (Akerlof and Shiller 2009, 13). However, under conditions of uncertainty, it is impossible to define precisely those options and probabilities. So in reality people do not act rationally but act according to what they trust to be true (rational). This also means that the decisions of economic actors depend largely on their beliefs and trust. Akerlof and Shiller emphasise the large role of confidence, for example, in the growth or decline of credit markets, and remark that the meaning of a term *confidence* goes beyond the rational and it is related with human feelings or, in other words, "animal spirits".

9.3.2 The Behaviour That Seems Irrational Can Lead to Rational Results

Behavioural economists, such as e.g. Gerd Gigerenzer, Daniel Kahneman, Amos Tversky and many others, point to the role of techniques, different from logical reasoning, which help people to solve problems and make the best decisions quickly. They emphasize the role of effort and time which a person must devote to rational analysis. Mental activity is associated with a high-energy consumption (the brain absorbs the most energy), so people, aiming at minimizing energy expenditure, apply heuristics – so-called "quick" thinking instead of logically analysing the problem ("slow" thinking), which claims less costs (calculated by time and effort).[3] Heuristic is a technique associated with a simplified way of thinking, a simple way to make a conclusion without resorting to mathematical calculations or scientific thinking. As Gerd Gigerenzer (2008, 20) underlines, "unlike statistical optimization procedures, heuristics do not try to optimize (i.e., find the best solution), but rather satisfice (i.e., find a good-enough solution)"; the models of heuristic cognition focus on situations in which people need to act fast. There are many heuristics, such as

[3] Kahneman (2012).

e.g. a rule of thumb, trial and error method, "imitate the majority", "imitate the successful" and so on. Daniel Kahneman (2012) discusses particularly the availability heuristics, heuristics of representativeness (used in the situation of assessing statements referring to probabilities) and heuristics of anchor and matching (it is used for the quantity evaluation). People use different heuristics depending on the situation and environment, and the same heuristic can be successful or not depending on the circumstances. In general, the use of heuristics can be explained by a diversity of human capabilities. As Gigerenzer notes, "without the evolved capacities, heuristics could not do their job". Among other examples he mentions the human capacity for recognition memory (such as face, voice, and name recognition), the capacity to imitate and the evolved capacity for reciprocal altruism (2008, 25).

It can be seen that the heuristics method is connected with human abilities that go beyond the limits of conscious activity, and reveals the richness of human nature, an important part of which consists of emotions (sympathy, antipathy, affection, fear, confidence and so on). In particular, with regard to long-term rationality, the emotional part of human nature is sometimes capable to challenge the boundaries of individual rationality based on hedonistic understanding of maximization utility. As an illustration imagine a greedy man who tries to "row for himself," i.e. to take more from other people than to give them. In the end, he risks losing friends and their possible support in case of need, etc. A rational person, who is capable of understanding the far-reaching consequences of his actions, can consciously be generous toward his friends (i.e. "invest" in his friends). These are two different motives for behaviour, but here it is possible to see how the specifically human needs and capacities (the need for friendship, the ability to be generous and kind) may on the outside be compared to the behaviour of a man whose rationality extends over a long period of time.

In the same vein, we can perceive the relation of human beings to the natural environment. Aside from people who do not pollute the natural environment only because there are external (public) prohibitions or because of pure economic reasons (in accordance with a logic of economic imperialists), there are people who do not pollute and do not damage nature because they feel affection for it and regard it as a living being. Therefore, the human love of nature contributes to the long-term preservation of the natural habitat, much like the rational actions based on the complex mathematical calculation of the future consequences of the present damage of nature would.

There are many other examples like this, e.g. in the realm of education and the investment in human capital, the creation of family, pension contributions, etc. But all of them in one way or another demonstrate that the bounded rationality and bounded human nature (in the sense of poverty of a human nature) converge at a certain point. In other words, the diversity of human nature compensates the limitations of human mind related to the lack of information and the inability to make long-term calculations. It seems in fact that to some extent, the rational and emotional parts of human nature substitute and enhance each other.

The Rational Behaviour of Individuals Can Lead to Irrational Outcomes In their book mentioned above, Akerlof and Shiller describe situations when, as it happened before the Recession of 2001 and the Great Recession of 2007–2008, individuals in financial markets behaved in accordance with the theory of rational choice, as they were following their own self-interests (2009, 35). However, since they invested in risky financial assets, their "rationality" did not lead to a macroeconomic equilibrium, but to speculative bubbles and eventually to a financial crisis. Therefore, in the end, their behaviour was not rational from the perspective of the economy as whole, and eventually from the perspective of the "rational" investors themselves.

9.4 Information, Complexity and the Principle of Emergence

Is a donkey rational, going after the carrot which is hanging on a stick in front of his muzzle? Having seen a carrot, a donkey theoretically can have a reason to take a step toward it. It can theoretically take a few more steps. We cannot call irrational the first steps of the animal, because the donkey is driven by hope, optimism and the absence of experience. If the donkey is stubborn in its hope, then we can call him the Donkey. The general conclusion of this example: the economic subject behaves irrationally, when he does not use (he does not try to use) accumulated experience. In other words, he does not accumulate information (his information resource equals zero), does not use it, i.e. does not learn and is characterized by (perfect) ignorance. In this example, the problem of bounded rationality, as it has been formulated by Simon, is not so much the problem of limited information (the static problem) but it must be connected with the problem of accumulation of information and the process of learning (dynamic context).

Accumulation (conservation and transmission) of information is a characteristic of complex systems. Complexity can also be defined by other considerations, including those of time and space. The economic system with a shorter lifetime is simpler in comparison with a relevant system with a longer lifetime. Thus, rationality associated with a short time utility maximization must be characterized as atomistic and also must be a characteristic of a very simple (socioeconomic) system. Since the length of time in question is connected with complexity, it is not accidental that, for example, in experimental economics the results for one-shot games (compare them with a short-term system) differ from results for repeated games (compare them with a more complex long-time system). Namely, according to the principal findings of experimental economics, in repeated personal, social, and economic exchange, as studied in two-person games, cooperation exceeds the prediction of traditional game theory (Smith 2008). In light of the foregoing, since a socioeconomic system has a high level of complexity and, correspondingly, of diversity and heterogeneity, one of possible answers on the question about irreducibility of macro (systemic) rationality toward individual one, may lie in the principle of emergence. The latter, in turn, means that "the whole is greater than the sum of its parts", i.e. that a complex system as an entity demonstrates properties that are absent in the constituents of the system, due to interactions among these constituent parts.

Furthermore, assuming the connection between rationality and information ability of a system, and also considering a bigger information ability of a more complex system, we can suppose that individual rationality is determined by collective rationality rather than a system (collective) rationality is being determined by the individual one. In general, it is consonant with the fact that a human as a reasonable being is possible only as a product of society ("social animal"). As Friedrich Hayek wrote, "[The] interaction of individuals, possessing different knowledge and different views, is what constitutes the life of thought. The growth of reason is a social process based on the existence of such differences" (Hodgson 2015, 292).

It is also consonant with the thesis (the result of research) that biological species with a high level of cooperation have a greater propensity to learn, adapt, survive, and, in general, to evolutionary development. The researchers point out that the cultural evolutionary process depends crucially on the size and interconnectedness of our populations and social networks; it is the ability to freely exchange information that accelerates adaptive cultural evolution, and creates innovation, and, at the population level, it is much better to be social than to be smart (Henrich 2015).

In other words, among other things, the rationality of individuals depends on their ability to learn from each other and from experience (to use accumulated information). Such a view seems to be similar to the approach of evolutionary and complexity economics. In particular, Alan Kirman remarks that, in markets, "the habits and relationships that people have developed over time seem to correspond much more to things learnt by the force of experience rather than to conscious calculation", and an attribution of rationality of the agents, when they are electing a strategy, is that "they are more likely to do what has proved to be successful in the past" (2010, 92, 85).

The inspiring examples of the superiority of collective rationality over individual rationality are given by natural sciences, in particular by behavioural ecology. For example, according to the research of Susan Edwards and Stephen Pratt, ant colonies can avoid irrational changes in preference that can be shown by individual animals and humans (Williams 2009). Edward and Pratt tested for irrationality in colonies of Temnothorax ants choosing between two nest sites that varied in attributes, such that neither nest site was clearly superior. In similar situations, individual animals show irrational changes in preference when a third relatively unattractive option is introduced. These societies act as unitary decision-makers, able to jointly select a single travel direction, foraging location or nest site from many options. Detailed analysis of this species has shown how consensus depends on a minority of active ants that scout for potential homes and assess their quality. Problem solving by insect societies relies on highly decentralized information processing. This partly reflects cognitive and information-processing constraints: individual insects cannot handle these problems alone, and colonies lack the hierarchical structures that might foster centralized decision-making. The results of this study support another advantage: the filtering out of systemic errors that would otherwise arise from the cognitive limitations of individual animals.

The example with ants echoes Hayek's idea cited by a number of well-known economists[4] according to whom the information in the economy remains dispersed and is never brought together into signals available to everyone:

> The problem of a rational economic order is determined precisely by the fact that the knowledge of the circumstances of which we must make use never exists in concentrated or integrated form, but solely as the dispersed bits of incomplete and frequently contradictory knowledge which all the separate individuals possess". "The whole acts as one market, not because any of its members survey the whole field, but because their limited individual fields of vision sufficiently overlap so that through many intermediaries the relevant information is communicated to all (1945, 519; 526).

One of the main features of a complex system is diversity, which, in turn, is related with information capability of a system. So, first, according to the information theory, the homogeneous structure is characterized by zero information. Second, according to W. Ross Ashby's law of requisite variety, to be able to resist variable unpredictable impacts of the external environment and thus to be conserved, the open system must have the requisite variety and complexity of its internal structure; only variety absorbs variety.

To sum up, on the one hand, if a social system consists primarily of "atoms" (i.e. selfish maximisers), its collective rationality is more reducible to atomistic rationality and, at the same time, such a system is more mechanistic, very vulnerable to external factors (sources of energy), and less enduring. On the other hand, in the case of a weak reducibility (or irreducibility) of systemic rationality to individual one, the system under consideration is more complex, it is characterized by more information capacity and diversity, and is more enduring.

9.5 Complexity and Institutions: Institutions as the Embodiment of Collective Rationality

An example of the influence of collective rationality on individual rationality can be the role that is played by institutions and norms. After all, in socioeconomic systems, it is institutions and culture that fulfil the function of preservation and conservation of information, i.e. they serve as the fundamental carriers and transmitters of information. Just as a human body "knows" how to function due to the information contained at the level of genes (genetic code), human beings often know what to do because they follow the norms and rules established in a society.

The link between institutions and information is emphasized by a number of prominent economists. For example, in the opinion of Douglass North (1991), institutions are formed to reduce uncertainty in human exchange with the help of structuring everyday life; they serve as indicators for human interactions; under conditions of incomplete information and imperfect computing capabilities,

[4] See, for example, Hodgson (2015, 292), Kirman (2010, 12), Smith (2008).

(institutional) constraints reduce costs of human interaction in comparison with the absence of institutions; in a world of instrumental rationality and complete information institutions are unnecessary. Also Geoffrey Hodgson (1988) points to the informative role of institutions and routines, in particular, when he writes that institutions really create and broadly disseminate additional information already by the very fact of their existence.

To better understand how institutions and rules predetermine the individual rationality, let us consider the following thought experiment. Imagine a magical external observer who watches people crossing the road. Let us also assume that he does not see the traffic lights (and does not know about their existence). And assume that there are no violators of the traffic rules. Let us consider the two following scenarios: first, when the traffic light works and, the second, when the traffic light does not work. So, in the first case, our observer can see the perfect order and think about perfect rationality of the both pedestrians and drivers: the pedestrians are rational, because they cross the road when the cars stand; the drivers are rational, because they stop when pedestrians cross the road. In the second case, however, the picture changes and becomes more chaotic: people can cross the street even when vehicles are moving; the cars can continue to drive even when people are crossing the road. Therefore, our magical observer can conclude that pedestrians as well as drivers became less rational. Is it really the case? In their own eyes, the pedestrians conserved both: a goal (end) – to cross a road, and a means – a wish to conserve their lives avoiding cars. The drivers conserved their goal to continue their way and the movement, avoiding pedestrians. In other words, their individual rationality did not change (or changed slightly). But something happened with the general system of rules, and the participants of the process under consideration were disoriented.

This thought experiment demonstrates, first, that individual rationality is not identical with collective rationality, and, second, that rationality of individual behaviour depends on the system of rules in a society. In other words, we can take pedestrians and drivers to represent economic actors, and a traffic light can be thought a representative of institutions, rules and routines dominating in the system under consideration. But, as Richard Langlois (1998) remarks, "economic choice as we normally think of it can happen only in a stable and predictable world in which most of the cognitive load is being carried by rules and routines". So the change or the damage of old institutions and rules influence the character of individual choice, which under conditions of growing uncertainty seems to become less rational.

It is a commonplace fact that culture, norms and institutions influence, for example, the tastes and, thus, decisions of consumers. Similarly, the institutional environment influences investment decisions. It is noteworthy that Akerlof and Shiller, pointing out the weak sides of the conventional theories of saving (which are constructed around the assumption about individual rationality), remark that "saving is largely cued by different institutional and mental frames" (2009, 123) and, in particular, give examples of the big differences in savings between China and United States, which are connected to institutional and cultural differences.

Actually, the phenomenon of individual rationality under the influence of the institutional environment is described by the concept of ecological rationality. The term of ecological rationality was coined by Gerd Gigerenzer (2008) and it is also used by Vernon Smith, the main creator of experimental economics, who considers that there are two types of rationality – constructivist and ecological, which coexist and complement each other. The concept of constructivist rationality is associated with the conscious deductive process of human reason, and the use of reason to deliberately create rules of action, and to create human socioeconomic institutions. Exactly this type of rationality is close to the understanding of rationality by neoclassical economics. The ecological rationality, in turn, is associated with intelligence embodied in the rules, norms and institutions of our cultural and biological heritage that are created from human interactions, but not by deliberate human design. In particular, one of the principal findings of experimental economics is that, with repeated experience in a variety of market institutions, impersonal exchange in markets converges to the equilibrium states implied by economic theory, even under information conditions far weaker than specified in theory (Smith 2008).

Thus, based on the foregoing, the question arises about conceptualizing institutions as the embodiment of collective rationality (or irrationality), which, in turn, affects individual rationality. With that we arrive at a hypothesis identical with Alan Kirman's assumption, namely that "it is the type of organization rather than the individual behaviour that is central to achieving coordination. Once in place, the organizational structure, itself, coordinates individual activities and makes them consistent" (2010, 6). In other words, in the interrelationship between individual and collective rationality, we have to start not from individual rationality in its neoclassical meaning, but vice versa.

9.6 Institutions and Human Emotions

Institutions are not only the product of constructivist rationality. As a form of human interactions, alongside with other factors, they can either build on certain human emotions or exploit those emotions, which, in turn, provide (or promote) human interactions. In other words, suppose that one of the most important components, which is built into collective rationality, but is excluded by individual rationality in its neoclassical understanding, is the emotional part of human nature.

In contrast to the selfish utility-maximizing model of an individual in mainstream economics, representatives of economic heterodoxy (such as e.g. evolutionary and anthropological economics) point to the altruistic and cooperative features of human nature, which are due both to genetic and cultural human evolution, and which are important for survival of the individual and the social groups (Hodgson 2015, 68–69). They emphasise the role of emotions for our social existence and note that "in a complex culture, emotionally empowered rules can help enhance notions of justice and morality". In particular, Geoffrey Hodgson notes that these features of human nature such as emotional capacities evolved by natural selection are "particularly important for the functioning of law and the state" (Hodgson 2015, 72–73).

Unlike representatives of economic imperialism who expand the principle of individual economic rationality on other, non-market, human relationships and attempt to present the altruistic and cooperative behaviour of human beings in the light of the logic of *homo economicus*,[5] i.e. treatise altruism as another form of self-interest,[6] the supporters of the complexity approach argue that both motivations for self-interest and generosity coexist, as obligation coexists with freedom (Cedrini and Novarese 2014).

Indeed, the fundamental socioeconomic and political institutions, such as, for example, institutions of democracy and market, rely both on the constructivist individual rationality and the emotional component of human nature related with non-selfish behaviour. For example, the capability of human beings to trust in others is one of the fundamentals of market transactions. According to different researches, this capability played a significant role in the human evolution, as it helped human beings to coexist together and to use advantages of cooperation and labour division. Thus, although markets are traditionally associated with competition and the search for personal gain, the characteristics of human nature such as adherence to moral principles and capability to trust also play an important role in securing the functioning of markets. In particular, Vernon Smith (2008), referring to the ideas of Adam Smith and his *Theory of Moral Sentiments* (1759), points to the fact of dependency of markets on human virtues and notes that otherwise the costs of monitoring and enforcement would become unbearable.

In their discussion of "animal spirits", alongside with confidence Robert Shiller and George Akerlof, among others, pay much attention to such human feelings as fairness and faith in stories. In particular, they note with irony that though some textbooks "do mention fairness as a motive, they still demote it to end-of-chapter, back-of-the-book status", and "it is reserved for those sections that student know they can skip when studying for the exam" (2009, 20). Such little attention to fairness seems paradoxical if we are to take into account, for example, what place Adam Smith devoted to considering "the sense of justice" in his *The Theory of Moral Sentiments* and that, in particular, he concluded:

> Beneficence is an ornament that makes the building more beautiful, not the foundation that holds it up; so it's good that it should be recommended, but it doesn't have to be imposed. In contrast with that, justice is the main pillar that holds up the entire building. If it is removed, the whole of human society must in a moment crumble into atoms (Smith 1759).[7]

To sum up, not only a human capacity of rational reasoning but a variety (complexity) of human nature in general, including its emotional part, has the impact on collective rationality since it influences the formation and functioning of socioeconomic institutions, whereas the latter, as it was demonstrated above, should be considered as a specific embodiment (or accumulators) of collective rationality. Thus, it makes sense to think of some key concepts and institutions of a market economy,

[5] See, for example, Becker (1974).

[6] The critique of this approach see, for example, in Cedrini and Novarese (2014).

[7] See Part II, Section 2, Chap. 3.

which are traditionally connected with neoclassical rationality, from the point of view of emotions embedded in these institutions.

9.7 The Institution of Innovative Entrepreneurship

Without taking into account the complexity of human nature, it is impossible, in particular, to explain adequately the phenomenon of innovative entrepreneurship as a driving force of economic development (in the understanding of Josef Schumpeter (2017)). For example, according to William Baumol,"the efforts of entrepreneurs are reallocated by shifts in the sectors of the economy and the lines of activity where profit seems most easily to be earned" (1993, 13). In our opinion, this is an important thesis that explains the intersectoral capital flows, and it is a financial investor who is highly sensible to profits (if to compare him with economic actors in the so-called real sector of economy). So, it is one of the basic motives for the financial capitalist (we use this controversial term here for convenience). As his task is "money making", he is (at least theoretically) indifferent about what to invest money as long as it is profitable. In other words, he is indifferent whether to invest money in the production of computers or in the production of slippers, or in financial speculations. Here we do not deny the importance of the financial investor in entrepreneurial activity, mindful of Schumpeter's thesis about the connection between the capital market and the development of the economy. But is Baumol's thesis true for the Schumpeterian entrepreneur-innovator who drives technological progress?

We dare suppose that the motivation of the great entrepreneur-innovators is much more complex, and the phenomenon of entrepreneurship cannot be explained exclusively by "easiness of the profit earning" and (or) profit maximization principle. For many of these personages we must acknowledge the combinative role of rationality and emotions (as e.g. a propensity to risk in the part not connected with mathematical calculations of probabilistic outcomes). For example, Henry Ford who undoubtedly was one of the greatest entrepreneurs in the industrial age said that "the highest use of capital is not to make more money, but to make money to do more for the betterment of life", and that "a business that makes nothing but money is a poor business". Steve Jobs, whose name is associated with the computer revolution, said: "Being the richest man in the cemetery doesn't matter to me. Going to bed at night saying we've done something wonderful, that's what matters to me"; "Your work is going to fill a large part of your life, and the only way to be truly satisfied is to do what you believe is great work. And the only way to do great work is to love what you do". He also acknowledged: "I have a great respect for incremental improvement, and I've done that sort of thing in my life, but I've always been attracted to the more revolutionary changes. I don't know why. Because they're harder. They're much more stressful emotionally. And you usually go through a period where everybody tells you that you've completely failed".[8]

It is noteworthy that Richard Langlois, when considering the question of rationality in relation to the entrepreneur-innovator who is dealing with novel situations (i.e. makes decisions in conditions of uncertainty), quotes Schumpeter's words: "Here the success of everything depends on intuition, the capacity of seeing things in a way which afterwards proves to be true, even though it cannot be established at the moment, and of grasping the essential fact, discarding the unessential, even though one can give no account of the principles by which this is done".[9]

In their *Animal Spirits*, George Akerlof and Robert Shiller also note that "the future of any country is in the hands of the business-people who decide on investments, and it is in large measure dependent on their psychology"; "business – at least successful business – thrives on excitement of *creating* the future" (2009, 143–144).

9.8 The Prisoner's Dilemma, the "Shadow of the Future" and Institutionalization of Emotions

The important question concerns the issue about the neoclassical type of individual rationality in its relation with socioeconomic institutions. If we start moving from micro rationality in the neoclassical sense, we risk concluding with the impossibility of cooperation and the viability of social institutions. One of the best known examples is the prisoner's dilemma, since it illustrates that individually rational behaviour does not necessarily lead to a socially optimal outcome.

The prisoner's dilemma, the originators of which are Merrill Flood, Melvin Dresher and Albert William Tucker,[10] is a standard example of a game analysed in game theory. It shows why two completely rational individuals might not cooperate, even if it appears that it is in their best interests to do so. Imagine that two members of a criminal gang – A (I) and B (II) – are arrested and imprisoned. Each prisoner is in solitary confinement with no means of communicating with the other. Each prisoner is given the opportunity either to betray the other by testifying that the other committed the crime ('defecting strategy'), or to cooperate with the other by remaining silent. The offer is: if A and B each betray the other, each of them serves 2 years in prison; if A betrays B but B remains silent, A will be set free and B will serve 3 years in prison (and vice versa); if A and B both remain silent, both of them will only serve 1 year in prison (on the lesser charge):

[8] https://www.brainyquote.com/quotes

[9] Schumpeter (1934/2017); Langlois (1998).

[10] See, for example, Tucker (1983).

		II	
		Cooperation	Defection
I	Cooperation	(1, 1)	(3, 0)
	Defection	(0, 3)	(2, 2)

This hypothetical scenario demonstrates that a choice to betray is a dominant strategy because defection always results in a better payoff than cooperation regardless of the other player's choice. Mutual defection is the only strong Nash equilibrium in the game (i.e. the only outcome from which each player could only do worse by unilaterally changing strategy). The dilemma, then, is that mutual cooperation yields a better outcome than mutual defection but is not the rational outcome because the choice to cooperate, from a self-interested perspective, is irrational.

In fact, the prisoner's dilemma deals with the perfectly atomistic approach to rationality as it ignores institutional environment and regards two prisoners who are members of the same criminal gang in a way that the fact of existence of this criminal gang does not influence the choice of the prisoners, and they do not have a fear of punishment from other members of this gang. Such rationality resembles rather the reflexive reaction of an animal on a piece of food in front of its muzzle, but not a work of human mind and reasoning. Is it not the case that the almighty mind of individual rationality, implying unlimited knowledge and possession of information, turns out to be only a primitive reflex of the animality greedy to the pleasures?

Moreover, in a such a type of interaction (a game), the assumption that each player is self-interested and always chooses the largest of two immediate payoffs for himself, strangely resembles the principle of entropy increase: if we consider the evolution of an isolated system, this unstable system left on its own will be destroyed, gradually converting into more probable and stable states; at the same time both probability and entropy are growing (Brillouin 1964). Since entropy is associated with a disorder in a system, the above-mentioned analogy inspires us to doubt the rationality as it is presented by neoclassical economics, and to think about rationality in a wider context, namely in the terms of the process of ordering and system complexity.

Also, the prisoner's dilemma does not presume emotional affection and trust (which can counteract the defective strategy) between persons because it does not take into account their past interaction, and also assumes that two individuals are destined never to meet again. In this situation, "no matter what the other does, the selfish choice of defection yields a higher payoff than cooperation" (Axelrod and Hamilton 1981, 1391). In their seminal article *The Evolution of Cooperation*, Robert Axelrod and William Hamilton (1981) note that, in many biological settings, the same two individuals may meet more than once. According to their model, probability of cooperation (correspondingly, probability of a defection strategy, but in opposite direction) depends on "the history of interaction so far" and the probability of the event, that after current interaction the same two individuals (players) will meet again. The latter is also expressed as "the shadow of the future" that must be long; no player should know when the game will end (Axelrod 1984).

For evolution of cooperation it is very important that "an individual must not be able to get away with defecting without the other individual being able to retaliate effectively"; for this it is necessary that "the defecting individual must not be lost in an anonymous sea of others". Axelrod and Hamilton (1981) note that higher organisms avoid this problem by their well-developed ability to recognize many different individuals of other species (they have a more complex memory, more complex processing of information; in humans, a better ability to distinguish between different individuals is largely based on the recognition of faces).

Based on game theory and the ideas of Robert Axelrod, it could be interesting to conceive of some institutions from the point of view of their role in the formation and support of "the shadow of the future" and, thus, promoting cooperative (non-defective) behaviour. For example, it can be the institution of church, i.e. the institutionalization of those human emotions and feelings, which are connected with the fear of death, faith in God (gods), faith in the afterlife, fear of punishment for sins (e.g. faith in karma). In other words, the church-supported faith in the afterlife and punishments from God, prolonging the "the shadow of the future", could facilitate the evolution of social cooperation; and this is one of the most obvious examples of how emotions are built into institutions, forming collective rationality (or irrationality).

9.9 Concluding Remarks

Rationality relates to information (knowledge), its accumulation and use. Thus, rationality is a characteristic of the complex, primary living systems. A feature of complex (living) systems is an accumulation of (free) energy and information, so the fundamental characteristic of rationality is to prevent the process of disordering and to the growth of entropy in a system. In this sense, rationality is an action (phenomenon or feature) that is intended to counteract the entropy processes and the growth of chaos in the system, and, thus, has similar functions to institutions, as the goal of both should be organization in a relevant system. In turn, neoclassical rationality in its connection with the idea of optimality and efficiency must be regarded as a particular case of this general anti-entropic foundation of rationality.

In the process of production of collective information, not only the human capacity to reason but also other human abilities such as risk appetite and the search for the new are involved, since they increment collective experience. Not only self-interested calculations of the future outcomes, but also human emotions such as feelings of affection, confidence, and fairness play an important role in collective coexistence and thus, influence social interactions which adopt the form of different institutions, and influence the formation and character of collective rationality. Collective rationality is embodied in social institutions and cannot be reduced to individual rationality in its narrow neoclassical meaning.

According to modern social psychology and neuropsychology, decisions based on emotions differ from decisions based on reason, but they help to support macro-rationality as they allow us to consider the interests of others in our decisions. So-called economic imperialism which extends the principle of individual economic rationality to other human relationships, not connected directly with economic activity, ignores the complexity of human nature, the role of human emotions and altruism, and overvalues the calculating capacities of the human brain. The emergence and functioning of many important socioeconomic institutions cannot be explained by narrowly selfish understanding of individual rationality.

A person is able to act both selfishly and altruistically and the way in which he/she will eventually act depends on the environment. A society which has much more information than an individual provides specific values that then become criteria for the future decisions of individuals. Political processes, like democracy sustained by free media, help the society to distinguish crucial values and to find and define the problems that the given society wants to solve. In its turn, the realized and expressed 'will' of a society takes a form of relevant institutions which then not only provide the criteria of 'be rational' and efficient but also have instruments to prevent or enforce that society to act according to these criteria.

In this sense, if the existing institutions favour altruism and not only egoism, there will be much more space and opportunities for individuals to develop their altruistic traits, especially if altruistic behaviour is considered by a society as rational. The neoliberal ideology, through its formal and non-formal institutions, expands the conviction that altruistic behaviour is not rational. At the same time, behavioural studies and social psychology open the new view, namely that altruism can pay off and in the end be a rational strategy. Societies with developed democracies seemed to have learned that diversity allows them to act more rationally as the whole and instil the values of the whole in individuals through relevant institutions. Diversity allows the society to minimize risks in an insecure environment and combat the challenges with which we have to deal.

Human greed and human generosity, egoism and altruism, reason and emotionality are embedded in different activities that can equally be important for the existence of human society as a whole. However, the prevalence of the sole characteristic, pushing out all the others, leads, in Ortega y Gasset's words, to pernicious homogeneity. So, starting from a narrow self-interest individualism, positivism and short-termism, we risk getting the "one-dimensional man" (the term used by Herbert Marcuse) who similarly to José Ortega y Gasset's "mass man" (Ortega y Gasset 1930), is capable of destroying the human civilization, including the intelligence that is the basis of human rationality.

References

Akerlof, G.A., and R.J. Shiller. 2009. *Animal spirits*. Princeton/Oxford: Princeton University Press.

Axelrod, R. 1984. *The evolution of cooperation*. New York: Basic Books.

Axelrod, R., and W. Hamilton. 1981. The evolution of cooperation. *Science* 211 (4489): 1390–1396.

Baumol, W. 1993. *Entrepreneurship, management, and the structure of payoffs*. Cambridge, MA: MIT Press.

Becker, G.S. 1974. A theory of social interactions. *The Journal of Political Economy* 82 (6): 1063–1093.

Brillouin, L. 1964. *Scientific uncertainty and information*. New York: Academic.

Cedrini, M., and M. Novarese. 2014. Economia, Altruismo e Dono. In *Il dono. Valore di legame e valori umani. Un dialogo interdisciplinare*, ed. G. Faldetta and S. Labate, 169–189. Palermo: Di Girolamo.

Gigerenzer, G. 2008. Why heuristics work. *Perspectives on Psychological Science* 3 (1): 20–29.

Hayek, F. 1945. The use of knowledge in society. *The American Economic Review* 35 (4): 519–530.

Henrich, J. 2015. *The secret of our success: How culture is driving human evolution, domesticating our species, and making us smarter*. Princeton: Princeton University Press.

Hodgson, G.M. 1988. *Economics and institutions: A manifesto for a modern institutional economics*. Cambridge: Polity Press.

———. 2015. *Conceptualizing capitalism: Institutions, evolution, future*. Chicago: University of Chicago Press.

Hogan, B., and L. Marcelle. 2017. The complementarity of means and ends: Putnam, pragmatism, and the problem of economic rationality. *Graduate Faculty Philosophy Journal* 38 (2): 401–428.

Kahneman, D. 2011/2012. *Thinking, fast and slow*. London: Penguin Books.

Kirman, A. 2010. *Complex economics: Individual and collective rationality*. London/New York: Routledge.

Langlois, R.N. 1998. Rule-following, expertise, and rationality: A new behavioral economics? In *Rationality in economics: Alternative perspective*, ed. K. Dennis, 55–78. New York: Kluwer Academic Publishers.

North, D.C. 1991. *Institutions, institutional change and economic performance*. St Louis: Washington University.

Ortega y Gasset, J. 1930/1993. *The revolt of the masses*. New York/London: W.W. Norton & Company.

Schumpeter, J. 1934/2017. *The theory of economic development: An inquiry into profits, capital, credit, interest, and the business cycle*. New York: Routledge.

Sen, A. 2002. *Rationality and freedom*. Cambridge, MA: The Belknap Press of Harvard University Press.

Simon, H.A. 1955. A behavioral model of rational choice. *The Quarterly Journal of Economics* 69 (1): 99–118.

———. 1997. *Administrative behavior: A study of decision-making processes in administrative organisations*. 4th ed. New York: Free Press.

Smith, A. 1759. *The theory of moral sentiments and on the origins of languages*, ed. D. Stewart. https://oll.libertyfund.org/titles/smith-the-theory-of-moral-sentiments-and-on-the-origins-of-languages-stewart-ed

Smith, V. 2008. *Rationality in economics: Constructivist and ecological forms*. New York: Cambridge University Press.

Sunder, S. 2002. Market as artifact: Aggregate efficiency from zero intelligence traders. In *Models of a man: Essays in memory of Herbert A. Simon*. Cambridge, MA: MIT Press.
Tucker, A.W. 1983. The mathematics of Tucker: A sampler. A two-person dilemma: The Prisoner's dilemma. *The Two-Year College Mathematics Journal* 14 (3): 228–232.
Walsh, V. 2007. Amartya Sen on rationality and freedom. *Science & Society* 71 (1): 59–83.
Williams, N. 2009. Collective rationality. *Current Biology* 19 (15). https://doi.org/10.1016/j.cub.2009.07.045.

Chapter 10
Naturalisation of Normative Economics

Marcin Gorazda

Abstract It is almost taken for granted among economists that the ultimate goal of public policy is to provide society with more "welfare", and the concept of welfare as well as the general strategy how to improve it is roughly the subject of studies of normative economics.

The naturalisation of normative economics is an attempt to analyse the "ultimate goal of public policy" from the perspective of human cultural evolution. The "ultimate goal" is one of the "cultural variants" which may be subject to Darwinian analysis and may or may not be adaptive in given circumstances. The most important problem which is posed here is whether such analysis could be helpful in establishing that "ultimate goal" and thus to contribute to normative economics and resolve the "normative problem" (as it is rephrased in the philosophy of law).

Three fundamental issues which seem to constitute the strong limits of the naturalization of the normative problem are highlighted:

1. Naturalistic ontology does not overcome the "naturalistic fallacy" indicated by Hume and Moore.
2. The epistemological perspectives in the naturalistic accounts are confused. Different normative postulates may be formulated from the individual perspective and from the perspective of the respective group, while the Darwinian analysis seems to privilege the population.
3. The genetic algorithm with endogenous fitness function seems to be non-susceptible to "mathematical close-up". Even if we know the initial normative order, there are limits to finding a shortcut in order to predict the future value of the fitness function. In other words, the predictability of the future social order is fundamentally restricted.

M. Gorazda (✉)
Copernicus Center for Interdisciplinary Studies, Jagiellonian University, Krakow, Poland
e-mail: marcin.gorazda@gsw.com.pl

P. Róna et al. (eds.), *Words, Objects and Events in Economics*, Virtues and Economics 6, https://doi.org/10.1007/978-3-030-52673-3_10

10.1 Introduction

Daniel Hausman (Hausman et al. 2017) in his recently published interview gave the following answer to the question about the ultimate goal of public policy:

> What is often said, which I think perhaps is justifiable, but not very helpful, is that a central goal of a government should be to promote the general welfare or general wellbeing. The reason I don't think that it is very helpful is that I don't think we have a good grip on what general wellbeing is.

This response perfectly reflects economists' first and spontaneous intuition of what is going on in normative economics. It is about the general wellbeing, and the main problem to be resolved is what it means.

Alfred Pigou, the indisputable father of welfare economics, was of a similar opinion. Although he did not see economics as a normative science, he claimed that it is "knowledge for the healing that knowledge may help to bring" and the most important task for economists is "to make more easy practical measures to promote welfare" (Pigou 1920, 30). He also provided us with the important insight of what welfare could be, looking for its foundations in the states of consciousness and their relations. However since policymakers have no access to the states of consciousness, and can hardly influence them, they need a convenient proxy which is money and economic welfare. The relation between money and welfare is not direct but is mediated through desires and aversions. Money does not measure the satisfaction received from the things it buys, but rather, the strength of our desires for those things (Hausman et al. 2017, 38).

Therefore Philippe Mongine (2002, 145) proposes the following definition of normative economics:

> The task of normative economics is to investigate methods and criteria for evaluating the relative desirability of economic states of affairs.

It sounds neutral (especially the phrase "relative desirability") but soon evoking Pigou, he instantly and directly refers to welfare economics, presenting four consecutive stages of its development (new welfare economics, social choice theory, modern welfarism). He also quotes eight basic assumptions of welfare economics:

I. Normative economics is an exclusive teleological theory which attempts at answering question about social good.
II. Social good is social welfare.
III. Social welfare is determined by the data of individual welfare.
IV. It exploits a particular notion of the social state which is determined by economic variables, primarily quantities of commodities consumed.
V. Individual welfare can be measured by an index of preference satisfaction.
VI. The index summarises individual choice behaviour (revealed preference theory).
VII. The index has standard properties of an ordinal utility function.
VIII. The index is not comparable from one individual to another (Mongine 2002, 160).

The assumptions are not indisputable, and in the course of the normative economics development some of them were undermined. Nonetheless, they provide us with a general concept of welfare as it is understood by economists. The concept reveals a strongly individualistic approach (with many reservations). The value of the social welfare function is supposed to determine the social good, while the ultimate social good revealed thereby determines the direction of public policy.

The idea that a somewhat ambiguous concept of general welfare should be the central goal of public policy is quite recent. It goes back directly to utilitarian ethics and partially to enlightenment ideology. Partially, because when we study the text of the three oldest constitutions (the American, the French and the Polish), definitely inspired by the European enlightenment, we will indeed find references to wellbeing, but this is not their central goal. In the American constitution, before welfare is mentioned, it reads: "more perfect union, justice, domestic tranquillity and common defence". In the French Declaration of Rights of Man and Citizen: "natural and imprescriptible rights of man" are put forward and "these rights are liberty, property, security, and resistance to oppression". And in the Polish constitution, we even find a passage which sounds very counter-welfaristic: "...holding dearer than life, than personal happiness the political existence, external independence and internal liberty of the people ...". The deeper we go into the human history and further from the European culture, the fewer references we find. Societies have a variety of concepts of their destiny and their ultimate goals. Throughout history and the world, welfarism is probably the least popular, while what seems to dominate are various theological concepts where the ultimate goal is subordinated to certain transcendental values (sacred natural rights, God's will etc.) or "tribal" concepts where it is defined in terms of a nation, state, tribe or tribal culture. We may reasonably conclude that:

1. Wellbeing understood roughly as described above by Mongine is far from an obvious answer about the ultimate social goal, regardless even of its ambiguity;
2. There might be some "natural" determinants of our concept of the ultimate social goal, and investigating them could be an interesting approach.

This paper attempts to investigate these possible, natural determinants of the concept of the ultimate social goal, primarily on the basis of the naturalistic, evolutionary approach. Assuming a certain mechanism of biological and cultural evolution, it tries to establish whether that mechanism might be instructive in the search for the social goal. In the first two sections, the problem is rephrased in terms of legal philosophy (the so-called "normative problem"), and the naturalistic jurisprudential approach is discussed. In the following section various evolutionary theories are presented, and then used to reconstruct the mechanism of the emergence of a normative order. In the closing sections two types of conclusions derived from the mechanism are drawn: Firstly, that in the light of the evolutionary mechanism, the present concept – the pursuit of wellbeing – may be maladaptive. Secondly, that the normative problem itself may be undecidable.

10.2 The Normative Problem in Naturalised Jurisprudence

Economics is not the only science which asks about the ultimate social goal. Similar question is asked by legal philosophers and the history of this question in this part of humanities is much longer than the history of normative economics. In jurisprudence, the question of the ultimate social goal, or in other words, the question of what law should govern the society, is often called "the normative problem" (Załuski 2014).[1] Since this phrase uses the same adjective as "normative economics", it will be applied further in the paper.

There are several currents in legal philosophy which have worked out their distinctive solutions to the normative problem. The most prominent among them are various sorts of natural law theories (some of them directly gave rise to the values evoked in the constitutions mentioned above), legal positivism and normativism, and legal realism. Since we will be looking for the "natural determinants", which roughly means scientific, and has nothing in common with natural law theories, we will focus on the last current. To naturalise the normative problem means to search for a solution within the science which describes human behaviour most accurately and comprehensively. Legal realism which was especially vivid in the USA in the first decades of the twentieth century developed an instrumental approach to law. Law was considered to be exclusively a means to an end.[2] Early legal realists, under the great influence of Benthamite utilitarianism, either did not notice that the concept of an end is not granted forever, or they thought that it could be established on the basis of scientific, natural studies of the regularities in human behaviour. Thus, they wanted to naturalise it. Both assumptions failed. The instrumental view of law found its disastrous culmination in the twentieth-century totalitarian regimes while the development of social sciences was too poor to draw sound conclusions. But the idea came back unsurprisingly with the progress in normative economics in the form of the Economic Analysis of Law, and with the progress in behavioural sciences in the form of experimental and evolutionary jurisprudence.[3] Those trends are the contemporary successors of the American realists, and they have also attempted to solve the normative problem.[4]

Before we elaborate further on the evolutionary approach, which seems to represent the most comprehensive theory of human behaviour and therefore is most promising, one additional remark on normative philosophy is in order. The sentences which are applied in ethics and law are not logical predicates, which means that they cannot be attributed truth value. They express duties and rights. They are the so-called "ought-sentences", and they require a different kind of logic than

[1] The normative problem should not be confused with the problem of normativity, which is also commonly discussed in the theory of law. The latter refers to the justificatory power of rules which makes us follow them and has nothing in common with the ultimate end problem.

[2] See Dewey (2008, 251) and Tamanaha (2006).

[3] See Beutel (1957).

[4] More on this in Gorazda (2017a, b).

classical predicate calculus. Whether such logic can be effectively constructed to conduct sound reasoning is an entirely different story. Lawyers are taught that to the certain extent it can, and they habitually perform such reasonings on the basis of a few inferential rules. A possible answer to the normative problem by its nature must also be the ought-sentence. Naturalisation of the normative problem means that its solution is sought in the natural facts (facts established in the natural sciences). From the logical point of view, it requires a special kind of reasoning: from predicates (factual sentences) to ought-sentences, in other words, from facts to duties. It is the so-called Hume/Moore problem as both philosophers agreed that such reasoning is not possible.[5] There are several proposals on how to solve this presumed impossibility, beginning from its denial,[6] through the extended kind of logic and inferential rules (Brożek 2013) and finally to the "modest naturalistic programme". The last one was advocated among others by Quine and is currently supported by Churchland. Briefly, it does not deny the genuineness of the Hume/Moore problem, but it states that even if at least one ought-sentence is irreducible to facts and must be necessarily accepted as a temporary axiom to conduct any sound deontological inferences, there is still enough room to reason on the specific duties. On the other hand, those specific duties are not inferred top-down (i.e. first general ultimate rule and then its derivatives) but, on the contrary, the usual way of constructing the normative order is bottom-up, i.e. first we resolve the specific cases, and then the applied, repetitive pattern of that resolving becomes the upper-level-rule.[7]

10.3 Evolutionary Philosophy of Law

Załuski specifies a few methods of naturalisation which are based on or at least refer to the Darwinian theory of evolution. Historically, the first to be mentioned is social Darwinism. The term is often used in a pejorative sense, and the concept which lies behind it is attributed (disputably) among others to H. Spencer. If the main 'goal' of evolution is 'the survival of the fittest' (a phrase coined by Spencer [1864, 444]), the ultimate goal of the order designed by humans should be the same. Otherwise, neither the order nor humans who designed it will survive. Social Darwinism was strongly and commonly criticised, mainly due to the dubious objectives of its supporters. But the most obvious weakness is the necessary pre-assumed and previous knowledge on the features which make someone the fittest. Another proposal listed by Załuski is "functionalism". If we could, in accordance with the evolutionary principles, decode the functions of the given psychological or behavioural mechanism, we could also design a normative order compliant to those functions. The similar logic can be reversed. Normative order should not require from humans that

[5] See Hume (2000), Moore (2004), Brożek (2013), and Hohol (2015).

[6] Cf. Searle (1964) and Pigden (1991).

[7] Cf. Quine (1979) and Churchland (1986).

they do what in light of the theory of evolution proves to be impossible or prohibitively costly. Those costs come from our predilections which are hardwired into human nature, which we can decode and thus conclude that any attempt at altering them will most probably fail. We owe this negative reasoning to legal theorist Owen D. Jones (Załuski 2018, 102). A conceptual tool inspired by evolutionary psychology has been worked out on the basis of this concept to assess the comparative effectiveness of legal regulations: the so-called "law of law's leverage". The possibility to determine certain 'function' of human behavioural traits or predilections opens the room for the determination of human nature in general or the direction of human evolution. Both can also be instructive for public policy and may give rise to "evolutionary ethics". Such reasoning led one of the most famous contemporary legal philosophers, H.L.A. Hart to propose the minimum content of natural law which is based on the assumption that certain rules are necessary for the realization of the 'minimum purpose of survival which men have in associating with each other' (1994, 193). Hart reconstructed certain features which may undermine our survival and then set against them rules, which, when obeyed, may neutralise those features. He listed five such features and their respective counteracting rules (Hart 1994, 195–196)[8]:

1. Human vulnerability to bodily attack, which is counteracted by the normative restriction of violence.
2. Human approximate equality in 'physical strength, agility, and intellectual capacities', so that in longer term nobody could effectively subordinate other members of the tribe, which entails rules constituting a 'system of mutual forbearance and compromise'.
3. Human limited altruism makes rules of mutual forbearance necessary and possible.
4. Limited resources for humans – are counteracted by 'some minimal form of the institution of property (though not necessarily individual property), and the distinctive kind of rule which requires respect for it'.
5. Human weakness of will and limited understanding of its long-term interest, which is counteracted by the system of sanctions which applies when certain rules are not observed.

But the most comprehensive, evolutionary theory of law we owe to Załuski (2009). In his account, knowledge of our biological and cultural evolution serves the determination of the most important features of human nature. Nature decoded thereby is compared to the concept of nature described in the most popular philosophical-political theories. As those theories usually include some factual assumptions (e.g. on the deemed human nature) and some normative judgments (e.g. on the most demanded content of social order) which are usually compliant with one another, the most accurate normative judgements are those made by the theory that is most consistent with the evolutionarily shaped human nature.

[8] Cit. after Załuski (2018, 63).

It is indeed true, that various philosophical-political theories imply a certain system of values which are promoted within such a theory. It is also true that we can learn a lot about our nature (defined after Załuski as the dominant moral motive and the way of conduct) by studying human evolution and prevailing daily practices. What may raise doubts is whether nature defined in this way exists and if it is stable enough to let us derive any conclusive views as to how would it look like in the foreseeable future, and if it is possible to derive any normative implications from such knowledge of that nature. Załuski claims that both questions should be answered positively. In reference to the former, there are accounts, rejected by the author, e.g., the so-called standard model of social science which claims that no such thing as stable human nature exists. Humans are more accurately modelled, as born *tabula rasa*, which is later shaped during an agent's path of life under environmental and cultural impact. This account, however, contradicts our evolutionary knowledge as well as contemporary empirical studies which broadly confirm that humans indeed reveal very repeatable patterns of behaviour. In Załuski's terms, they are narrowly altruistic (kin and reciprocal altruism is commonly observed) and imperfectly prudent. If we agree on the existence of human nature, to combine this notion with normative judgments, theoretically there are three stances possible. The weak version agrees that the correct recognition of human nature is necessary for an effective introduction of any social policy, but it gives up searching for any ultimate goals of the law. The weaker and more general version also agrees that the ultimate goal cannot be derived from human nature, but its recognition has an impact on social practices. The strongest version (supported by the author) goes furthest and claims that stable human nature determines the goals of the law by the philosophical-political theory consistent with it and supported by it. Such a theory usually has two components which should be compatible with one another. These are the concept of human nature and the set of value judgements. If the concept of human nature embedded in the theory coincides with the knowledge constructed on the basis of the evolutionary approach and empirical studies, we have strong reasons to believe that the embedded value judgments are also sound. In the conclusions, Załuski notes that the narrowly altruistic and imperfectly prudent human nature can hardly be compliant with communism, anarchism, conservatism, and libertarianism but does not contradict liberalism and (to a lesser degree) socialism.

Do we really have reasons to assume that human nature does resemble the one pictured above by Załuski? Although his remarks are supported by some empirical studies, especially those investigating the contemporary patterns of human behaviour, there are also significant grounds to question them. Those patterns are not stable enough, and they seem to evolve even in the historical period. This observation was raised by several authors studying the evolution of human culture and its impact on our behavioural patterns (Richerson and Boyd 2005). Particularly interesting are the studies on the declining propensity towards violence in societies which coincide with the studies on cooperation in larger societies. This somehow counterintuitive and surprising conclusion was among others expounded by social and evolutionary psychologist, Steven Pinker (2011). Pinker's starting point is that our neurobiological mechanisms make us far from being gentle or friendly towards each other. On the contrary, our "hardware" is rather programmed to various forms of violence. Paradoxically, even those cooperative dispositions, which are recognised by Załuski as features supporting collaboration in society and the creation of legal order, may be responsible for ideological subordination and violent intergroup rivalry. Kin and reciprocal altruism do strengthen social ties but within ideological groups, thus making them more inclined to aggression towards other groups.[9] Pinker also notes that we are equipped with some mechanisms (both biological and social) which let us tame our natural tendency towards aggression. These are empathy (bound with our narrow altruism), self-control, moral sense (in Załuski's terms: personal and moral autonomy) and intelligence. It is worth noting that some of these features may be responsible both for aggression and its curbing as it is in the case of the above-mentioned cooperative dispositions fuelled by empathy, or moral sense which may steer our behaviour to "justified" but violent revenge. Given such a complex picture of several counteractive mechanisms, the real question is not about the stable features of human nature but rather why the latter, responsible for curbing violence, has prevailed since the fact that they prevail is broadly confirmed by thorough and numerous empirical studies. After analysing some candidates for the explanation and rejecting the implausible ones (like genotype evolution in recent years), he proposes five causes, all of them of cultural character. In the first place, he puts Leviathan, i.e. the organised state and its institutions. This state's priority strongly resembles the account of Hobbes, rejected by Załuski. The second factor is the development of commerce which forces merchants to take the position of their customers if they want to cut any deal with them and thus it strengthens cooperative behaviour. In giving primacy to commerce, Pinker is not alone. He subscribes himself to the Austrian school of economics and its successors. Many of the contemporary studies on human evolution which aim at explaining our dominance over other humanoids also point to human propensity to exchange goods, even among very distant tribes. The third factor is feminisation which curbs the harmful impact of testosterone and male rivalry. The fourth is the extending circle of sympathy. In order to break through the narrow, family or tribal empathy, we need to frequently meet strangers in peaceful circumstances, talk to them and get more familiar with their perspective and emotions. This process accelerated with the growing geographical and social mobility, and with widening literacy and the custom of reading books and stories coming from various parts of the world and cultures. This is directly connected to the last factor, "the escalator of reason." Although analytical, cognitive systems are used for many forms of violence, in the end, the reason is more likely to tame it. Reason means more self-control and higher intelligence, both interdependent. Since psychologists have started to measure intelligence, they noticed the so-called Flynn effect – an observed increase of IQ in each generation. This strengthens the pacifying impact of reason. More intelligent people are inclined

[9] This evolutionary ambivalence of human nature has also been noted by Załuski in his later works. See Załuski (2018).

to offer collaboration instead of instrumental violence; they are more liberal, agreeable for group decision-making and constructive, gentle discourse.

Pinker's particular theses have been criticised by many academics from a different perspective. However, at least two of them seem to be somewhat untouched: the complex picture of human nature, rather biologically predisposed to violence than to friendly co-operation; and the changing patterns of human behaviours over the centuries. Both theses undermine Załuski's optimistic view. Clearly, it is the other way around. The moral and legal order is not a simple extension of our biological predispositions but rather helps us to curb them. And our nature observed statistically in repeatable patterns of behaviour is an effect of a subtle game between several counteractive biological mechanisms and social environment while both factors act interdependently. The same picture is drawn by several, evolutionary oriented cultural anthropologists like Richerson, Boyd and Henrich. They discovered, among others, the particular mechanism of co-evolution of genes and culture which was responsible for shaping the human social instinct, which further gave rise to large, modern cooperative societies. Through the so-called "moralistic punishment" this evolutionary dance may gradually eliminate non-cooperative agents and thus modify our, statistically defined human nature. In the end, it seems that Załuski's concept does not provide us with a plausible solution to the normative problem.

10.4 Neglected Cultural Evolution and the Emergence of the Normative Order

Załuski's attempt to construct a unified philosophy of law based on our evolutionary shaped propensities is not the only one. If we initially agree that the normative order, within which the normative problem is posed and answered is *per se* an evolutionary phenomenon, which is somehow shaped by an agent's propensities, but even more by the influence of the prevailing culture, it still seems promising to track down the mechanism of its emergence. The comprehensive knowledge of this past mechanism may give us a hint about the future normative problem solution. Several theories were proposed to that effect. The most prominent were offered by Hayek, Quine and nowadays by Churchland, Boyd and Richerson. There are some differences among them but what they have in common is the bottom-up approach, spontaneity of the behavioural patterns emergence (driven by random forces like cultural mutation and cultural drift[10]), group selection mechanism (especially in Hayekian theory[11]) and thus the strong influence of culture and evolutionary pressure in the form of natural and sexual selection. In contemporary social sciences, the most comprehensive picture is drawn by Richerson and Boyd (2005) and in the philosophy of law by Brożek (2016), who widely exploits Tomasello's (1999) account. The

[10] Richerson and Boyd (2005, loc. 977).

[11] Hayek (1983).

common element in those accounts is an emphasis placed on culture, as the main normative-order-creating factor and the use of mathematical modelling which strengthens the presented arguments. Culture is understood as:

> ...information capable of affecting individuals' behaviour that they acquire from other members of their species through teaching, imitation and other forms of social transmission. (Richerson and Boyd 2005, loc.105)

Culture includes instructions or imperatives, which are cultural variants transmittable between agents, societies and generations. Those variants are subject to three categories of cultural evolutionary forces.

1. Random forces, which are on the agent's level, cultural mutations, and on the group level, a cultural drift.
2. Decision making forces, which by their nature are more consciously chosen by an agent, and consist of the guided variation (emulation) and biased transmission (cultural variants imitated from others due to their deemed effectiveness).
3. Natural selection, which is based on an unbiased transmission (cultural variants imitated naturally from parents and close relatives) and changes in the composition of a population, due to sexual selection and elimination of unfit agents.

Culture in general is a powerful evolutionary adaptation and cultural variants and genes co-evolve. However it may be, and sometimes indeed is maladaptive. Cultural evolution and the spread of cultural variants within a population can be tracked down and modelled by the Darwinian analysis, which means the extensive use of genetic algorithm and agent-based modelling.

A genetic algorithm was initially developed in informatics and mathematics to improve problem-solving tools. It resembles an algorithm which steers biological evolution and it assumes that if evolution is so good at solving an adaptation problem, similar instructions to solve less sophisticated technological problems should be used. However, besides solving problems, the genetic algorithms disclosed one more side-effect. It enhanced our understanding of the actual evolutionary process by creating an opportunity to simulate them and observe results depending on different values of variables applied.

How does a genetic algorithm work then? Like its evolutionary model, it makes use of similar terminology, namely population, chromosome, mutation, crossing over and offspring, and consists of several consecutive steps (Mitchell 1999, 308).

1. We start with a randomly generated population of *n l-bit* chromosomes (candidate solution to a problem), then
2. Calculate the fitness $f(x)$ of each chromosome x in the population.
3. Repeat the following steps until n offspring have been created:
   (a) Select a pair of parent chromosomes – probability of selection being an increasing function of fitness.
   (b) Cross over the pair at a randomly chosen point to form two offspring.
   (c) Mutate the two offspring at each locus with probability p_m and place the resulting chromosomes in the population.

(d) Replace the current population with the new one.
(e) Go to step a.

In the search for an ultimate goal, the second step seems to be most interesting. It assumes that in each task we have a predetermined fitness function which enables us to calculate the results which we intend to achieve and which we value as the most desirable. The functioning of the algorithm and the direction in which chromosomes evolve are subordinated to the value of the fitness function. If by analogy, we apply the same mechanism to human evolution, the search for an ultimate goal turns into a search for a proper fitness function and its optimum value. However, is it possible to predetermine it in a natural environment? In the simplified model above, the function is exogenous. In models imitating natural evolution, it must be endogenous, and it evolves depending on the initially embedded genome, the changing environment, and an increase in the population, and the last element is crucial. The more numerous are the carriers of a particular chromosome; the more likely this chromosome is to spread within the population. And *vice versa*, the more likely the chromosome is to contribute to the expansion of the population, the more likely it is to survive, spread, and represent the searched value of the fitness function. The changing environment complicates the whole story. The optimum value should remain optimum both in short and in the longer term. So, the sought chromosome is required not only to be stable but also flexible enough so that in the long term, the survival of the population is not threatened. If we anchor our reasoning in the fact that they were cultural changes which shaped our behavioural patterns to form a full-blooded legal order rather than our genome and, that it is a population as a whole which may determine the value of a fitness function rather than a particular agent's predispositions, it will lead us to the concept of cultural evolution through group selection.

The Hayekian theory was one of the first accounts based on the group selection mechanism, also referred to by Załuski. According to Hayek, controversial group selection is responsible for cultural evolution. Certain traits/patterns of behaviour could have evolved because they increased the fitness of the group, even if they decreased the fitness of an agent. The group is more fit if its chances of survival rise in a confrontation with other groups. However, there is no unique, distinguishable carrier of a cultural trait (memes). Traits can spread within the group (and be transmitted to other groups) in two ways. The first is an imitation. It makes cultural evolution Lamarckian by its nature, and not Darwinian. The second is sexual selection – the desired traits are more often chosen by sexual partners, which makes them more likely to be spread within the population. This would work if we accepted the concept of genetic group selection so that certain genetically encoded traits, which are not adaptive for an agent but adaptive for the whole group, may nevertheless spread in the way described above and dominate other traits. Excluding the very exceptional example of human lactose tolerance, no other case of genetic group selection has been recorded so far. That is why the account is controversial.[12]

[12] See for example Price (2011).

However, it is commonly accepted that learned skills may indeed be under evolutionary pressure connected with the group selection mechanism.[13] If so, the account is not complete. Darwinian evolution, which is the only one admitted as a scientific theory, does not allow the skills learned by phenotype over its lifetime to be passed on to the next generation. How is it possible then?

The first person to suggest a plausible answer was James Mark Baldwin, who in 1896 published a paper on the possible passing on of specific learned skills to the next generation. It is not the particular skills that are passed on but rather the cognitive flexibility of an agent (the trait which is genetically determined), which enhanced strongly an agent's ability to learn those skills during its lifetime. Those who are more gifted and learn quicker, live longer and are more likely to have more offspring, which explains the rapid spread of this cognitive ability in the population. Those abilities according to Brożek (2016) (and Tomasello, whose works he refers to) are imitation and emulation. Imitation means an elementary form of copying the behaviour of others, according to the rule "do what I do" without reflecting on the purpose of conduct. Emulation is more like the process of learning. It cannot be reduced to imitating other agents, but it requires awareness of the ultimate end of the process. I learn not to know but to achieve a certain, predetermined, ultimate goal. It means that with the use of my intelligence, I do not restrict myself to imitate others, but I am consciously able to decline in some points from the observed patterns and steer my behaviour towards the previously identified goal. The significant contribution of Tomasello is that he found out that whereas imitation is broadly observed among mammals, especially primates, emulation is specific for humans. But imitation also has some specifically human peculiarities. Although primates can imitate others like humans, it seems that their propensity to imitate is significantly weaker. As Brożek puts it, "(…) their tendency to imitate is limited in comparison to human eagerness to copy others. We do indeed 'out-ape' the apes, but not because they have no skill for aping – they lack the tendency" (Brożek 2016, 670). From this point on, Brożek's account significantly supplements the Baldwin effect. Referring to one of the simulation models of Boyd and Richerson, he claims that these two uniquely human features are the foundation for any, even very primitive, form of culture, a part of which is a normative order. "It transpires that in such a setting there exists 'the evolutionary equilibrium amount of imitation', from which both learning-prone and imitation-prone individuals benefit, so that the population as a whole has a higher average fitness than a population consisting solely of learners, or a population in which there are only "pure" learners and imitators. They further observe that "imitation may increase the average fitness of learners by allowing learned improvements to accumulate from one generation to the next" (Brożek 2016, 705). Most importantly, imitation is a culture-creating mechanism – it enables the transmission of behavioural patterns from generation to generation, thus allowing the accumulation of knowledge, smooth adaptation of existing cultural tools, the recombination

[13] See Richerson et al. (2016), where authors presented the set of arguments for cultural group selection being an essential mechanism in human evolution contributing among others to emergence and development of cooperation among humans.

of means and ends, as well as the emergence of fine-grained ways of conduct (Brożek 2016, 740). The essential element in this picture that should be added is that both learned skills inheritance paths (i.e., imitation and emulation) do not act evenly within the population. Some of the agents are more likely to be imitated or to be learned from. First and foremost, they are parents and close relatives.[14] Secondly, the outstanding individuals perceived as successful according to the commonly accepted standards[15] and, thirdly, others, while our propensity to imitate them declines with the increasing cultural distance, language being one of its most important indicators.[16] The author further draws before the reader the possible path from imitation and emulation through the emergence of rudimentary rules and their consecutive transformation into fully-grained abstract rules, expressed in a language; complex, modal and endowed with the justificatory power. From the subject issue of this paper, it is worth noting that the searched ultimate goal, the solution for the normative problem is always embedded in the abstract rules and thus, in the extended social order. To ponder over humans' desired destination, we need a language, modality and justification. Rudimentary rules do not need any predetermined ultimate end, or at least an agent does not need to be aware of it.

As we have some plausible hypothesis on the learned skills transmission mechanism, we may go back to the Baldwin effect and its possible simulation in models based on a genetic algorithm. Evolutionary Reinforcement Learning (ERL) model designed by Ackley and Littman (1992) is a simple one, but it includes all the essential components of cultural evolution. Agents in the model move through a two-dimensional space. As in the natural environment, they forage food, may encounter a predator and hide from it. They dispose of a certain amount of energy, which is consumed and partially recovered during their lifetime. Exhaustion of energy means an agent's termination. Their key equipment is two neuronal networks: one is responsible for the evaluation of the given state, whether it is good or harmful for an agent (evaluation network), and the other one is responsible for the undertaken action (action network). The first network may be interpreted as a surrogate of simple normative judgment. The architecture of the network is identical for each agent, but it differs in weights. As a consequence, they value their states differently. The genetically encoded chromosome determines the evaluation network and thus is subject to the consecutive steps of the genetic algorithm (mutation, cross-over, multiplication, selection, and replacement) and invariant during the lifetime. In contrast, the action network is subject to a learning process and thus could be modified during the lifetime. An agent learns to act in ways that lead to "better" states. "Better" means the initially inborn evaluation, then modified according to the survival rate.

[14] This effect is very well recognised and confirmed by several experimental studies. It is referred to by Longman (2004). See also Palmer and Steadman (1997). In Poland, research was among others conducted by Brzozowski (1988), Wołosiuk (2010) and Elżanowska (2012).

[15] See Henrich (2004), Henrich and Boyd (1998), Henrich and Gil-White (2001).

[16] Boyd calls this unbiased and biased transmission and they both, together with natural selection, constitute the three model factors responsible for spread and prevalence of cultural traits. Richerson and Boyd (2005, loc. 935).

What is most important, no exogenous fitness function is given for evaluating the genome. Fitness is endogenous; it emerges from many actions and interactions (and evolves). It means that no ultimate end is given from the outside. The measured parameter is the time of survival of the given population before it becomes extinct. It depends on the following initial settings: evolution plus learning (ERL), evolution alone (E), learning alone (L), (so that the evolutionary mechanisms of mutation, cross-over, multiplication, selection, and replacement are excluded), fixed weights chosen randomly (F) and random walk through the matrix, (i.e., ignoring the information from the environment) (B). The extension of the population and its survival time gives us some basic information on which setting is the most effective one and what combination of the value judgements (evaluation network) and learnt skills (action network) is the most adaptive and stable, and thus whether we may draw any conclusions on the possible ultimate goal, the ultimate system of value judgements for the initial population. After several iterations, Ackley and Littman were able to summarise their results. ERL did much better than evolution alone (E) and slightly better than learning alone (L). Fixed weights worked worst, even worse than random walk (B). "It is easier to generate a good evaluation function than a good action function" (Richerson and Boyd 2005, loc. 427). It is easier to specify useful goals (encoded in the evaluation network) than useful ways of accomplishing them or, in other words, the evolutionary mechanism works well at choosing the best value judgments but cannot cope effectively with selecting the appropriate means to accomplish these values. Working out an effective *modus operandi* requires the engagement of a learning process. The second part of the simulation additionally supported this interpretation. Both scientists decided to search for the so-called functionally constrained parts of a chromosome. The underlying concept is that in nature those parts of the genome which are subject to the least changes through generations are supposed to be the most functionally constrained ones, which means that they are most likely to be strongly adaptive. To check which parts could those be in the model, they extended the lifespan of the population to almost 9 million generations. It occurred that genes responsible for encoding the evaluation network reveals a low level of variation during the first 600,000 generations until the chromosome reaches the optimum pattern for the action network. Then this setting becomes more functionally constrained, loosening the constraints on the evaluation network. So it seems to be more important for shaping the proper ability for an agent's current state evaluation in the early stages of the population's evolution, which is necessary to strengthen the learning ability. Stable goals are crucial for survival. However, later on, when proper modes of action begin to be genetically encoded, value judgments play a less critical role.

If this simulation correctly imitates the milestones of natural biological and cultural human evolution, we may suspect that patterns of our behaviours are more important for our survival than value judgments including judgments on ultimate end, assuming the early stages of our evolution are behind us. It somehow coincides with the observation that the very foundational principles of human behaviours indeed remain invariant throughout the very distant cultures while the system of abstract rules (moral and legal), often involving deeper value judgments, presupposed ultimate human goals and its justification, differs significantly even within the same cultures and geographical areas.

Let us now consider the following conclusions which may be derived from the preceding passages. I have summed them up below being fully aware that they are intuitions or plausible guesses rather than a proven thesis. However, if they were correct, we might draw up the following picture:

1. Human nature, defined as a dominated moral motive and typical way of conduct, is hardly definable independently of time and place. Contrary to Lock and Załuski, it seems to be relatively unstable and variable throughout generations, and its underlying biological mechanism inclines us rather towards violence and aggression than to co-operation (which unfortunately supports the dismal picture drawn up by Hobbes).
2. The evolutionary mechanism has no fitness function. The spread of particular traits within the population is subordinated to its expansion and the strength of imitation and emulation or biased and unbiased transmission. In case of behavioural patterns transmitted through imitation and emulation, the expansion of the particular population (symbolically marked groups[17]) seems to be a stronger factor as we are more inclined to imitate our close relatives (unbiased transmission), and the spread of group-beneficial beliefs is twice as quick within the group as from one group to another (Richerson and Boyd 2005, loc. 2865).
3. Rudimentary rules and abstract rules may correspond to the action network and evaluation network in the Ackley and Littman model. The former is crucial for survival, the latter is less functionally constrained, at least after the adaptive patterns of behaviour are preserved enough.
4. Both create social orders at different levels. Abstract rules with their justificatory power include value judgments, and among them the desired postulates regarding ultimate goals.
5. Both social orders of different levels may interact with one another. Rudimentary rules give rise to abstract rules, but it goes the other way around, too. An extended order (system of abstract rules) may act in our way of conduct (rudimentary rules), thus causing, for example, the observed decline of violence.

The outlined picture gives us a relatively comprehensive (although necessarily simplified) image of the possible evolutionary origin of human normative order. It provides us also with suppositions for where to look for the ultimate end value judgments and the main factors shaping them. At the first stage, they must be genetically encoded and functionally constrained, but in the pace of a species' further development, they become less constrained, and their impact on our behaviour weakens, being reduced to interactions on the social order levels. In other words, we may have very different ideas on what is and what is not good for us, or where we should head (namely, very different value judgements), but the essential patterns of our behaviour remain relatively unchanged. It does not exclude possible changes, even

[17] See Richerson and Boyd (2005, loc. 2865).

commonly spread, but as they are caused mainly by the system of abstract rules and corresponding extended order, their causal power is more vague and subtle, yet it may also create certain observable trends such as the decline in violence.

10.5 The Pursuit of Wellbeing May Be Maladaptive

There are two significant consequences of the models presented above. Firstly, the value judgement and especially the determination of the ultimate end may be of less relevance to the evolutionary success of human species than we might have thought. What counts are our behavioural patterns, which at a certain stage of the species' development are weakly dependent on value judgements. The significant part of them may be of no relevance to our behavioural patterns. They may play a role of necessary chromosome mutations, while most of them have no significant impact on the phenotype or, even if they have, they may be maladaptive, and they may terminate. Secondly, the solution to the normative problem is not reducible to the search for the proper fitness function. In other words, evolution (both biological and cultural) is directionless. Any sort of 'social Darwinism', 'functionalism', 'evolutionary ethics' or 'minimum content of law' is bound to fail, especially if it is primarily dictated by the survival and wellbeing of an individual agent. The spread of behavioural patterns seems to be more strongly subordinated to the expansion of the population. Patterns which are more likely to contribute to that expansion are more probable to spread. If at the end some of them reveal any causal power, the "survival rate" is stronger among those contributing to the population's increase than those that contribute to the agent's wellbeing. This is one of the most important discoveries in evolutionary biology and the unexpected consequence of the population's thinking. Early evolutionary biologists thought that it was the survival and fitness of an individual that counts the most, but further research reveals that selection favours traits that increase the reproductive success of individuals and not their fitness. The same refers to cultural evolution and the emergence and diffusion of a particular normative order.

> Biased transmission depends on what is going on in the brains of imitators, but in most forms of natural selection, the fitness of different genes depends on their effect on survival and reproduction independent of human desires, choices and preferences. (Richerson and Boyd 2005, loc.1106)

This slightly counter-intuitive effect of evolution may, for example, explain the surprising spread of Christianity in the Roman Empire, despite the tough and puritan ethic which clashed with Roman promiscuity (Załuski 2012). It may also occur that subordination of women, which is still the case in some Islamic states, will have a higher survival rate only if it implies higher birth rates (which is plausible), despite strongly negative ethical connotations. It may at the end occur that the pursuit of wellbeing or individual happiness, which has been commonly preached recently, is counter expansive (maladaptive) if the pursuit of wellbeing or happiness merely means pursuit of comfort and avoidance of unpleasant and disturbing circumstances like childcare. This is precisely the kind of explanation offered by Richerson and Boyd of the surprisingly low fertility rate in western civilisation. However, for them, the reason does not lie primarily in the pursuit of comfort but rather in the maladaptive patterns (or as they name it, cultural variants). Biologists know that evolution is not a perfect process which always leads the species to maximise their fitness, but on the contrary, it is full of errors and maladaptive traits. A genetic algorithm is not able to eliminate those traits, especially when the same trait carries both adaptive and maladaptive functions. The often-invoked example is the peacock's tail. Its size is at the same time a visible sign of strength and health of its carrier, which makes it an attractive mate, and an obstacle when it comes to escaping predators, which puts its carrier at risk of premature termination. However, sexual selection seems to be a stronger factor so far, as males with a bigger tail have more offspring before they fall prey to the predator. Boyd and Richerson constructed a similar argument in reference to the so-called prestige bias. In principle, culture and specifically biased transmission of the cultural variants is very adaptive, as it helps us to work out and sustain the skills which are crucial for our survival, and it does it in a much shorter term than genes evolution would have done it otherwise. As it has already been demonstrated in the model mentioned above, a certain level of imitation of other people, especially our parents and prestigious members of society, is required to sustain the culture and to spread the cultural variants which contributed to the success of the prestigious ones. On the other hand, those who are commonly perceived as prestigious in a modern western culture very often reveals the "selfish cultural variant". It is quite obvious when we consider who the people are whom we perceive as successful. At least from the beginning of the industrial revolution, they are rich merchants and entrepreneurs, highly paid professionals, popular artists and scientists. To reach their social position they needed to sacrifice much of their time and resources for education and professional training. Those who follow them, but are not equally talented or endowed, need to sacrifice even more. High status also needs to be socially marked. In modern societies, those markers are often expensive toys and hobbies. On the other hand, childcare is costly and extremely time-consuming. Prestige bias forces drive us away from those costs. It is especially true for women. The deepest fall in the fertility rate is strongly correlated with their access to education.[18] The "runaway" cultural evolution is thus accelerated by the universal education and development of mass-media which "suddenly exposed people to much more non-parental culture influence than had been experienced in more traditional societies".[19] It is worth noting that this non-parental culture is at the same time the culture promoting individual wellbeing, understood as preference satisfaction and the pursuit of individual happiness, which happens to be the necessary components of rapidly spreading "selfish cultural variant", apparently maladaptive.

[18] See Richerson and Boyd (2005, loc. 2467) and Newson and Richerson (2009).

[19] Richerson and Boyd (2005, loc. 2365).

10.6 Concluding Remarks: Threefold Undecidability of the Normative Problem

Does it mean that humans will become extinct? Does it mean that the concept of wellbeing is nothing but a dangerous maladaptation which at a certain stage of our cultural evolution brought us some individual comfort but in the longer term may drive us to extinction? The truth is that no one knows. But we cannot exclude such a hypothesis. It seems that the normative problem is undecidable regarding the naturalistic approach for at least three independent reasons.

1. Naturalistic ontology does not overcome the "naturalistic fallacy" indicated by Hume and Moore. Even if we adopt the modest naturalistic programme and consequently assume certain indisputable norms as axioms (like the discussed concept of wellbeing), we instantly encounter three unsolvable problems:

   (a) According to the modest naturalistic programme, axioms are not created in the value-judgement vacuum and do not reflect our, indisputable human or divine intuition. They are constructed bottom-up, by numerous, previously resolved normative problems, solutions of which were subsequently evolutionarily tested. Axioms are generalisations reflecting our multigenerational knowledge and seem to be a more useful tool for solving daily normative problems than somehow privileged, higher-level rules.
   (b) The causal power of those axioms is doubtful. As Litman and Ackley's model shows, the higher-level value judgements, which are crucial at the early stages of species' evolution as they determine the behavioural patterns, soon after those patterns are genetically encoded, become less functionally constrained. At least some of them may serve as useful rationalisations of our behaviour, without having any real impact on them.
   (c) If some of those axioms reveal some causal power, the development of the social order according to them and according to the principles determined by the genetic algorithm may occur to be evolutionarily counterproductive, as Boyd and Richerson's model on fertility rate decline shows.

2. The epistemological perspectives in the naturalistic account are confused. Different normative postulates can be formulated from the individual's perspective and the perspective of the referential group. What seems to benefit an agent, needs not to be beneficial for the population, and the other way around, what is beneficial for the population, needs not to benefit an agent. As the endogenous fitness function seems to privilege the population, traits which are beneficial exclusively for agents and do not contribute to the genetical success may be condemned to extinction. The concept of wellbeing may represent such a trait.
3. The genetic algorithm with an endogenous fitness function seems not to be susceptible to "mathematical close-up". Even if we know the initial normative order, there are limits to finding a shortcut in order to predict the future value of the fitness function. In other words, the predictability of the future social order is

fundamentally restricted. Models based on the genetic algorithm are by their nature very complicated, but at the same time, they are an enormous simplification of the actual environment and forces acting behind it. They help us to understand certain mechanisms and trends but will not reveal to us the future state of society and especially the future evolutionary solution to the normative problem.

The third point is especially the direct reason for the undecidability of the normative problem, the reason why we cannot determine if the concept of wellbeing is an example of the maladaptive trait condemned to extinction. There are many scenarios possible. Extinction is one of them. Richerson and Boyd point to particular orthodox groups like Methodists and Amish in the USA, who have been able to cut themselves off from the access to the modern mass-media and thus weaken the biased transmission and cultivate their traditional family and religious values, which implies high fertility rate. Wellbeing, especially one that is understood in monetary terms, is definitely not their ultimate end. They are slowly growing in numbers. On the other side of the social structure, we have the growing model of lonely parenting. The fertility rate among those who apply this model is not very high but at least close to the natural replacement rate. It may be that a certain equilibrium between the counter-acting processes like the spread of selfish cultural variant, fertility rate and natural selection will be reached in the longer term. Finally, there is the problem of "estimating the strength of various effects on the trajectory of evolution".[20] There seem to be natural and robust reasons to reject moral Darwinism. As the fitness function does not exist, and the interdependence between the different social orders is multi-causal, any possible value of adaptiveness is untractable.

Taking into account the above-outlined picture of the possible origin of the normative social order and the remarks on its potential consequences, it seems that in our search of the ultimate end, we are within a vicious circle with no way out. It appears that we are condemned to accept the fact that the ultimate end problem is undecidable, at least within the evolutionary knowledge.

Not everything is lost, however. At least, following the "modest naturalistic program" we may form a couple of reasonable postulates:

1. In social sciences (including normative economics) "nothing makes sense except in the light of evolution". Darwinian analysis and populational thinking should be an integral part of those sciences. Commonsensical truths, which are taken for granted are nothing more than cultural variants which may and should be analysed in terms of their adaptiveness (welfare, happiness, success, pleasure and like).
2. Predictability of the possible consequences followed by the adoption of certain cultural variants is limited but not entirely excluded. Big data collection and agent-based models combined with the Darwinian analysis may help draw possible scenarios and be informative for societies and policymakers.

[20] Richerson and Boyd (2005, loc. 3486).

3. If moral axioms have a weaker impact on our behavioural patterns (which can be the subject of possible, future empirical studies), than we have suspected, their relaxation may not constitute a real danger for social cohesion. And if they play a role of necessary cultural mutations, wherefrom both adaptive and maladaptive patterns may emerge, which will be naturally tested and observed, their diversity may be desirable, but only from the particular point of view, namely the survival and further expansion of the human population. This could be interpreted as an argument for a liberal and open society and value pluralism, as it used to be promoted by K.R. Popper (2013) and I. Berlin (2002).

References

Ackley, D. & Littman, M. 1992. Interactions between learning and evolution. In *Artficial life II*. s.l, ed. C.G. Langton, C. Taylor, J.D. Farmer, and S. Rasmussen, 487–509. Redwood City: Addison-Wesley.

Baldwin, J.M. 1896. A new factor in evolution. *American Naturalist* 30 (441–451): 536–553.

Berlin, I. 2002. Two concepts of liberty. In *Liberty*, ed. H. Hardy, 166–217. Oxford: Oxford University Press.

Beutel, F.K. 1957. *Some potentialities of experimental jurisprudence as a new branch of social science*. Lincoln: The University of Nebrasca Press.

Brożek, B. 2013. Błąd naturalistyczny z perspektywy logicznej. In *Fenomen normatywności*, ed. B. Brożek, A. Brożek, J. Stelmach. 179–194. Kraków: Copernicus Center Press.

———. 2016. Imitation and the emergence of normative orders. In *The emergence of normative order*, ed. J. Stelmach, B. Brożek, and Ł. Kurek, Kindle ed. Kraków: Copernicus Center Press.

Brzozowski, P. 1988. Przekazywanie wartości w rodzinie. *Psychologia Wychowawcza* 2: 190–199.

Churchland, P. 1986. *Neurophilosophy: Toward a unified science of the mind–brain*. Cambridge: MIT Press.

Dewey, J. 2008. *The middle works, 1899–1924. Volume 10 1916–1917*. London/Amsterdam: Feffer & Simons.

Elżanowska, H. 2012. Międzypokoleniowa transmisja wartości w rodzinie. *Studia z Psychologii w KUL* 18: 97–114.

Gorazda, M. 2017a. *Granice naturalizacji prawa*. Warszawa: Wolters Kluwer.

———. 2017b. Law as a means to an end. Undecidability of the end problem. *Polish Law Review* 3 (1): 68–84.

Hart, H.L.A. 1994. *The concept of law*. Oxford: Oxford University Press.

Hausman, D.M., M. Gorazda, and T. Kwarciński. 2017. O przydatności ekonomii, celach polityki publicznej i znaczeniu filozofii ekonomii. *Zarządzanie Publiczne* 42 (4): 102–109.

Hayek, F.A. 1983. *Law, legislation and liberty. Volume I. Rules and order*. Chicago: The University of Chicago Press.

Henrich, J. 2004. Demography and cultural evolution: How adaptive cultural processes can produce maladaptive losses: The Tasmanian case. *American Antiquity* 69 (2): 197–214.

Henrich, J., and R. Boyd. 1998. The evolution of conformist transmission and the emergence of between group differences. *Evolution and Human Behavior* 19: 215–242.

Henrich, J., and F.J. Gil-White. 2001. The evolution of prestige: Freely conferred deference as the mechanism for enhancing the benefits of cultural transmission. *Evolution and Human Behavior* 22 (3): 165–196.

Hohol, M. 2015. Naturalne podstawy normatywności: Hobbes i Hume. In *Naturalizm prawniczy. Stanowiska*, ed. J. Stelmach, B. Brożek, Ł. Kurek, and K. Eliasz, 22–33. Warszawa: Wolters Kluwer.
Hume, D. 2000. *A treatise of human nature*. Oxford: Oxford University Press.
Longman, P. 2004. *The empty cradle: How falling birthrates threaten world prosperity and what to do about it*. New York: Basic Books.
Mitchell, M. 1999. *An introduction to genetic algorithms*. Kindle ed. Cambridge/London: MIT Press.
Mongine, P. 2002. Is there progress in normative economics. In *Is there progress in economics*, ed. S. Boehm, C. Gehrke, H.D. Kurz, and R. Sturn, 145–170. Northhampton: Edward Elgar Publishing.
Moore, G.E. 2004. *Principia Ethica*. Mineola/New York: Dover Publications.
Newson, L., and P.J. Richerson. 2009. Why do people become modern? A Darwinian explanation. *Population and Development Review* 1 (35).
Palmer, C.T., and L.B. Steadman. 1997. Human kinship as a descendant-leaving strategy: A solution to an evolutionary puzzle. *Journal of Social and Evolutionary Systems* 20: 39–51.
Pigden, C. 1991. Naturalism. In *Companion to ethics*, ed. P. Singer, 421–431. Oxford: Blackwell Publishing.
Pigou, A.C. 1920. *The economics of welfare*. London: Mcmillan.
Pinker, S. 2011. *The better angels of our nature: Why violence has declined*. New York: Viking Books.
Popper, K.R. 2013. *The open society and its enemies*. Princeton: Princeton University Press.
Price, M. 2011. *The controversy of group selection theory*. http://www.scq.ubc.ca/the-controversy-of-group-selection-theory/. Accessed 27 Ferb 2012.
Quine, W.V.O. 1979. On the nature of moral values. *Critical Inquiry* 1 (3): 471–480.
Richerson, P.J., and R. Boyd. 2005. *Not by genes alone. How culture transformed human evolution*. Kindle ed. Chicago: The University of Chicago Press.
Richerson, P., et al. 2016. Cultural group selection plays an essential role in explaining human cooperation: A sketch of the evidence. *Behavioral and Brain Sciences* 39: 1–68.
Searle, J.R. 1964. How to derive "ought" from "is". *The Philosophical Review* 73 (1): 43–58.
Spencer, H. 1864. *Principles of biology*. London: Williams and Norgate.
Tamanaha, B.Z. 2006. *Law as a means to an end. Threat to the rule of law*. Cambridge: Cambidge University Press.
Tomasello, M. 1999. *The cultural origins of human cognition*. London: Harvard University Press.
Wołosiuk, B. 2010. Wpływ preferencji wartości rodziców na aksjologiczną orientację dzieci. In *Wartości w rodzinie: ciągłość i zmiana*, ed. W. Muszyński, 64–74. Toruń: Wydawnictwo Adam Marszałek.
Załuski, W. 2009. *Evolutionary theory and legal philosophy*. Cheltenham/Northampton: Edward Elgar Publishing.
———. 2012. O teorii kulturowego doboru grupowego na przykładzie analizy rozwoju chrześcijaństwa w cesarstwie rzymskim. *Logos i Ethos* 32.
———. 2014. *Game theory in jurisprudence*. Kraków: Copernicus Center Press.
———. 2018. *Law and evil: The evolutionary perspective*. Cheltenham/Northampton: Edward Elgar Publishing.

Chapter 11
Beyond Mere Utility-Maximisation. Towards an Axiologically Enriched Account of Well-being

Tomasz Kwarciński and Wojciech Załuski

Abstract The aim of the article is to reconstruct the way the concept of well-being is most commonly understood in economics, to identify deficiencies of this understanding, and to find a remedy to them. The paper defends two basic claims: (1) that the dominant understanding of well-being in economics is non-normative and boils down to the concept to utility maximisation, i.e., satisfaction of one's preferences, whatever they are (subject, at best, to some formal constraints); (2) and that, given the shortcomings of the non-normative concept of well-being, e.g., its tautological character (at least in the case of certain formulations of this concept), it is indispensable to build an axiologically richer (that is: normative) concept of well-being, thereby going beyond the mere notion of utility-maximisation. As for the claim (2), two versions of the normative concept are distinguished in the article, viz. an exclusive one (well-being as causally dependent on prudential values but not moral ones) and an inclusive one (well-being as causally dependent on prudential values constrained in some way by moral ones), and it is argued that the latter is more plausible, since, in contrast to the former, it does not rely on a dubious assumption about the absolute priority of prudence over morality.

11.1 Introduction

Somewhat schematically, one can say that there are three assumptions economics makes about well-being. The first one is that the economic agent is focused on what is "good for herself" (is self-interested). The second – which adds content to the first one – is that well-being is the primary criterion of evaluation of the state of affairs in normative economics: what is good for an agent equals her well-being. The third – a natural consequence of the second one – states that the economic progress

T. Kwarciński (✉)
Department of Philosophy, Cracow University of Economics, Krakow, Poland

W. Załuski
Department of Philosophy of Law and Legal Ethics, Jagiellonian University, Krakow, Poland

P. Róna et al. (eds.), *Words, Objects and Events in Economics*, Virtues and Economics 6, https://doi.org/10.1007/978-3-030-52673-3_11

of societies is evaluated in terms of social well-being usually measured by GDP *per capita*. As we can see, well-being is a central concept of economics. However, there are controversies as to how it should be understood. Our goal here will be to reconstruct the way the concept of well-being is most commonly understood in economics and to identify deficiencies of this understanding. More specifically, we shall argue that it is necessary to build an axiologically richer concept of well-being and that in order to do this, one has to go beyond the mere notion of utility-maximisation (which is usually assumed by economics to be *definiens* of well-being). If such understanding of well-being is accepted, the dichotomy of self-interested and axiological motivation of economic agents can be overcome.

The structure of the paper is as follows: Section 11.2 argues against a non-normative account of well-being which reduces this concept to utility maximisation, i.e., satisfaction of one's preferences, whatever they are (subject, at best, to some formal constraints). Section 11.3 discusses two versions of a normative account of well-being: (1) well-being as causally dependent on prudential values but not moral values (an exclusive approach); (2) well-being as causally dependent on prudential values constrained in some way by moral ones (an inclusive approach). The final section concludes.

11.2 Against the Non-normative Account of Well-being

The prevalent view of well-being in mainstream economics can be dubbed non-normative, as it refers merely to utility maximisation, where utility is defined either in terms of revealed or stated preferences. Before we discuss these two versions in some detail, let us notice that both of them constitute a significant departure from the classical understanding of utility. In his popular textbook "Intermediate Microeconomics", Hal Varian asserts that back in the Victorian days classical utilitarians and economists understood 'utility' as an equivalent of a person's overall well-being identified with happiness (Varian 2010). However, this claim is not entirely correct. In fact, Jeremy Bentham (1907) says that what is meant by 'utility' is "that property in any object, whereby it tends to produce benefit, advantage, pleasure, good, or happiness (all this in the present case comes to the same thing) or (what comes again to the same thing) to prevent the happening of mischief, pain, evil, or unhappiness to the party whose interest is considered." Thus, it would be more precise to say that, for him, 'utility' was a synonym of 'usefulness' (Broom 1999, 19), which can give rise to certain positive consequences (including happiness). But this is only a starting point in a complicated history of the meaning of 'utility' in economics. The meaning evolved from signifying a useful property of things (Bentham), through mental states of a person (pleasure, happiness), the intensity of his desires and their satisfaction (Pigou 1932, 38), to its contemporary understanding – as preference satisfaction among specified alternatives. What is more, even the relationship between utility and preference satisfaction changed. As Varian (2010, 55) states: "Originally, preferences were defined in terms of utility: to say a bundle (x1,x2) was preferred to a bundle (y1,y2) meant that the x-bundle had a higher utility than the y-bundle. But now we tend to think of things the other way around. The preferences of the consumer are the fundamental description useful for analysing choice, and utility is simply a way of describing preferences. A utility function is a way of assigning a number to every possible consumption bundle such that more-preferred bundles get assigned larger numbers than less-preferred bundles." The question of which consumption bundle is preferable to someone is answered upon the choice he or she makes. For instance, if someone consistently chooses apples when bananas are also available, an economist can infer that she prefers apples over bananas. Thus we have come to the first version of the non-normative account of well-being (in terms of revealed preferences) mentioned at the beginning of this section. In our critical analysis of this concept we shall follow, to a large extent, Amartya Sen (and we shall also use the term 'well-being' interchangeably with 'welfare' – the term Sen prefers).

The lack of a choice-independent way of understanding a person's attitudes towards alternatives provides a rationale for what is called definitional egoism which means that "(...) all agent's choices can be explained as the choosing of 'most preferred' alternatives with respect to a postulated preference relation" (Sen 1977, 323). Following Amartya Sen, we can identify three dimensions of self-interested behaviour. Firstly, this view assumes that a person's welfare depends only on his or her consumption bundle (self-centred welfare). Secondly, the only goal of a person is to maximise his or her welfare (self-welfare goal). Thirdly, people always act purposively, that is, whatever they are choosing it is directed to reach their goals according to their order of preferences (self-goal choice) (Sen 1986, 7; 1988, 80). The self-centred welfare precludes the possibility that an agent's welfare depends on the consumption bundles of some other persons. For instance, it rules out the possibility that someone feels worried because of another person's extreme poverty and that this feeling negatively impacts his own well-being. The self-welfare goal excludes any attachment of any positive or negative weights to other people's welfare from the agent's goals. The agent cannot aim to increase or diminish other people's welfare. Finally, the self-goal choice states that the only motive behind a person's actions is his or her desire to further their goals. All other motives, such as complying with social conventions or acting according to one's religious and moral commitments have to be excluded.

The combination of self-welfare goal and self-goal choice allows one to infer that an agent makes his or her choice exclusively to further his or her welfare, which means that she always prefers and chooses things which enhance her own welfare (self-welfare choice). The self-welfare choice establishes choice – welfare inference by ruling out choices unrelated to the agent's welfare, and not allowing the possibility of counter-preferential choices. The self-welfare choice, together with the self-centred welfare, completes the task of defining personal welfare on the basis of her choices. On this view, whatever one chooses, the choice will mean that it is her most preferred option and that it maximises her well-being. In other words, what is chosen by an agent is always good for her or him. Thus, definitional egoism, implied by the theory of revealed preferences, leads to the following tautological account of well-being:

> The life of an agent S goes well if and only if S succeeds to choose, in the overwhelming majority of decisional situations he is faced with, the options that are the most preferred for him; and the statement that an option p is most preferred to S in a given decisional situation means that S chooses p in this situation.

The arguments against this view can be divided into two groups. The first one consists in a criticism of the assumption that the self-goal choice is true. The second group contains arguments rejecting this assumption. According to Daniel Hausman, the choice – welfare inference is doubtful even if people always make purposive choices. He points out that in order to infer welfare from an agent's choices through her preferences, economists have to assume not only that the agent is goal-oriented and self-interested but also that she is aware of different available options, and can correctly judge what is better for her. Hausman gives the following example. "Suppose there are only two alternatives, x and y, and that x is in fact better for Jill than y. If Jill judges correctly, then Jill will rank the expected benefit of x above that of y. If Jill is self-interested, then Jill prefers x to y. If Jill knows that she can choose x or y, she will then choose x. The economist can then work backward. From Jill's choice, the economist can infer Jill's preference – but only on the assumption that Jill knows that she could have chosen y. From Jill's preference for x over y, the economist can infer that Jill thinks that x is better for her than y is – but only on the assumption that Jill is self-interested. From Jill's judgment that x is better for her than y is, the economist can conclude that x is, in fact, better for her – but only on the assumption that Jill's judgment is correct (Hausman 2011, 89)." Hausman's analysis shows that definitional egoism, in fact, excludes the possibility that the agent's choice could be mistaken. Given how often people are wrong about their knowledge regarding available options or correct judgements, this is a rather unconvincing assumption. Thus, Hausman concludes that a person's choices do not determine what is good for her but can only be treated as evidence of her well-being. Even more fundamental objections against definitional egoism are raised by Amartya Sen. According to him, choice cannot be a part of the definition of an agent's welfare because sometimes other people make choices for her. For instance, if the government decides how much money to spend on defence so that its citizens do not feel insecure, we cannot infer citizen's well-being levels merely by looking at their choices. What is more, going beyond self-goal choice, an agent can be motivated by other reasons than her welfare (even broadly understood). Someone can follow social rules or moral norms which are against his or her preference ordering. In such a case his or her choices cannot be seen as enhancing their welfare (Sen 1980, 206).

The above definition of well-being forms a kind of non-normative view because it does not contain any reference to values (prudential or moral). However, we should be aware that the rejection of definitional egoism does not necessarily lead to the normative interpretation of well-being. There is a second version of non-normative account, according to which:

> The life of an agent S goes well if and only if, in the overwhelming majority of decisional situations S is faced with, S succeeds to choose the options that are the most preferred for him; and the statement that an option p is most preferred to S in a given decisional situation means that it occupies the highest place in his ranking of preferences which is constructed or given prior to the choice.

Unlike the previous one, this account (in terms of stated preferences) avoids the objection of being tautological, because it assumes that preferences are known prior to making a choice. But it is still non-normative, as it implies that only formal requirements are imposed on an agent's ranking of preference. In particular, a set of preferences should be complete and transitive. Preference completeness means that an agent can either prefer x over y or y over x or is indifferent between them. It is essential to distinguish between indifference and inability to rank options. If an agent is indifferent between two options, it means that each of them is equally good for him, while if he is unable to rank them, not only does he not know whether one of them is better than the other but also whether they are equally good for him. The requirement of transitivity implies that if an agent prefers x over y, and y over z he also has to prefer x over z. This requirement protects the agent from being exploited, since, as the so-called "money pump argument" shows, a person with intransitive (cyclic) preferences will be eager to engage in a series of transactions where he "(...) will have paid a positive price for a zero benefit" (Schick 1986, 116). For instance, someone who prefers x over y, and y over z, and at the same time z over x will be ready to sell x to buy y, paying an extra money, then sell y to buy z, paying an extra money, then sell z to once again buy x, of course paying extra money. The series of transactions ends up at the point when an agent is left with the same good which he had at the beginning, but with less money. Although these formal requirements are important for an agent to ensure that his life will go well, this account of well-being, as we have seen, does not make any reference to prudential or moral values. So even though it is not a tautological concept of well-being, it is still non-normative: it does not rely on any substantive (axiological) assumptions. All of this leads us to the conclusion that in order to create a richer, normative account of well-being, we have to reject definitional egoism and move beyond purely formal constraints of preference ordering.

11.3 The Normative Account of Well-being

One can distinguish two ways to account for the normative conception of well-being: one in which well-being refers exclusively to prudential values, and the second in which besides prudential values well-being includes some reference to moral ones. Let us investigate them in turn.

11.3.1 The Exclusive Approach: Well-being as Unconstrained *Pursuit of Prudential Values*

The exclusive approach to normative well-being assumes that there is only one kind of value which contributes to a person's well-being, namely: prudential value. If something is prudentially valuable for someone, it means that it is good for him. The general theory of what it means that x is good for y was presented by Richard Kraut (2009). He claims that to say that x is good for y, we have to know the nature of both elements of this relationship as well as whether there is a certain fit or match between them. For instance, watering is good for potted flowers or changing the oil in a car is good for the engine or having a couple of devoted friends is good for a person. To say something about what is good for a person, that is what contributes to his or her well-being, we have to gain some knowledge about him or her.

As economists usually assume, the basic facts about human beings are a reflection of the basic tenet of folk psychology – that people have some beliefs and desires. Roughly speaking, these mental entities are what enables them to form goals and make decisions which are supposed to lead them to the attainment of these goals. More technically, persons are capable of forming preferences and want them to be fulfilled. According to this view, p is good for S if and only if p satisfies S's preference. For instance, if Eve would like to have a child and she, in fact, gives birth, then the baby is good for her. However, if she prefers not to have a child, and the reality matches her preferences, then not having the baby will be good for her. Because a person is an agent with a set of beliefs and desires, whatever is good for a person depends solely on her or his preferences. The match between a person's preference and her preferable state of affairs depends on the occurrence of that state which guarantees satisfaction of the preference.

This somewhat formalistic view of prudential values, however, is an object of wide criticism. First of all, since a person exists in particular circumstances, her preferences can be influenced by many factors, which leads to the well-known problem of preference adaptation. What is more, preferences can be formed via indoctrination; if they are formed in this way, they cannot be regarded as a genuine reflection of wants and beliefs of that particular person. One possible answer to these objections is to move into the direction of the theory of true preferences. On this theory, only preferences created by agents who are in the proper state of mind, that is, who always reason with the greatest possible care, and have all relevant information about the states of affairs should be taken into account (Harsanyi 1977, 646).

Finally, the view that something is good for a person – is prudentially valuable – just because she prefers it, may work well in some popular examples from economic textbooks but not necessarily in real-life situations. For instance, if someone has to choose between ice-cream and chocolate (a common dilemma in many economic textbooks), it is undoubtedly true that that dessert is good for him, which he prefers most. However, in real life, people often prefer something because they are convinced that it is good for them. Following James Griffin (1993), we indicate that prudential values require going beyond the so-called taste model (that is, the view that value of things is based only on personal tastes or preferences) towards the perception model which states that in order to prefer something, a person has to recognise its value first. In other words, people should prefer something because it is valuable, not the other way around. This way of conceptualising the relations between values and preferences is assumed in Griffin's account of prudential values. He lists these values under the following categories: "(1) accomplishment (the sort of achievement that gives life point and weight), (2) the components of a characteristically human existence (autonomy, liberty, and minimum material provision), (3) understanding, (4) enjoyment, and (5) deep personal relations" (Qizilbash 1996, 155; Griffin 1993, 52). Usually, people share these values, though not all of them are equally important for all persons. It may be the case that something good for one person can be bad for someone else. The important thing is that if something is suitable for a particular person, according to Kraut (2009) it fosters a flourishing of the person. It is worth noticing that even this richer view of prudential values (as compared with the previous one which defines values in terms of preferences) can conflict with moral obligations. It may be the case that taking care of other people's welfare, that is being altruistic, will require a sacrifice of some part of the agent's well-being (e.g. if I devote some of my time to help other people at the cost of not realizing the prudential value of enjoyment, I sacrifice some of my prudential well-being for the sake of moral values). Now, on the exclusive account of the normative well-being, such sacrifice always amounts to the diminution of well-being. This approach can be more precisely stated in the following way:

> The life of an agent S goes well if and only if (a) S's preference set contains a sufficiently large number of prudential preferences (i.e. preferences expressive of his attachment to prudential goods/values); (b) S chooses prudential goods in many decisional situations; (c) in the case of conflict between prudential values and moral values the agent always gives priority to the former.

As we have already mentioned, the normative force of this account of well-being rests on its reference to prudential values which are not purely subjective but depend on the perception of what is good for a human being. It is debatable whether this is an agent-neutral or agent-relative normativity; if the former, every person has a reason to promote or at least to not undermine other people's welfare; if the latter, a person has a reason to support only her own well-being. But we shall analyse this problem further. We shall focus instead on a different question, namely: whether (as it is assumed in the exclusive approach) well-being can indeed be achieved by unconstrained (by moral values/obligations) pursuit of prudential values. We shall argue that it cannot.

11.3.2 *The Inclusive Approach: Well-being as* Constrained *Pursuit of Prudential Values*

Let us recall that the exclusive approach to the normative account of well-being resolves the two main problems of the non-normative account. First, it is non-tautological, which means that the agent's action cannot always be (*ex-post*) interpreted as contributing to her own well-being. It is true that this problem has also been resolved within the second version of the non-normative account – but in an essentially different way: by imposing certain formal requirements on the agent's preferences and assuming that they are known *prior to* making a choice, and not (as in the exclusive approach) by introducing certain substantive requirements with regard to the content of the agent's preferences. Resolving this problem by recourse to the exclusive approach is much more effective: it is by far easier to state that an agent failed to reach the state of well-being if the notion of well-being is defined in a more restrictive manner – by allowing not only formal but also substantive requirements. But this methodological advantage, as one may call it, is less important than the second one. The basic problem of the non-normative account is that it seems to be an incorrect explication of the concept of well-being, for the concept, on its ordinary usage (though not on its usage that is dominant in economics!), has *a normative dimension*. Moreover, this dimension cannot be reduced to purely formal constraints imposed on the agent's preferences. The exclusive approach to the normative account undoubtedly does *some* justice to this dimension. It, therefore, also resolves (at least to some extent) the second problem. But the question arises whether it does it *full* justice, i.e., whether it does not present well-being in an overly reductionist manner. Let us deal with this question at somewhat greater length. It will lead us to a richer account of well-being, one that makes some reference to moral values.

The exclusive account assumes (correctly, in our view), that the essential (in a causal or constitutive sense) components of well-being are prudential values. If the agent does not pursue them, or if she pursues them unsuccessfully, she cannot be said to have achieved the state of well-being. This claim seems to be true in the empirical sense (prudential values do seem to be the basic source of the positive subjective state, which we tend to call 'well-being'). But in addition to this plausible claim, the exclusive account also includes a highly controversial one: that an agent who wants to achieve well-being should always give priority to prudential values in the case of their conflict with the moral ones. It is important to understand this (controversial) claim properly: it does not say that a prudential agent can never act morally. If it did, it would be evidently false because the demands of prudence are often convergent with those of morality (that "honesty is always the best policy" is, regrettably, false, but it is for sure true that "honesty is *often* the best policy"). What it does say is that prudence and morality may give rise to mutually inconsistent claims and that, if we want to achieve well-being, we should *always* sacrifice the demands of morality. This claim is very difficult to evaluate. It is not even clear what method of evaluation one would need to assume. Whether the method should be conceptual (one would then have to examine perhaps by 'questionnaire' methods of experimental philosophy, whether the concept of well-being, as it is commonly used, implies that absolute priority of prudential values over moral ones), or empirical (psychological research examining which of the two types of agents – those never sacrificing prudential values for the sake of the moral ones or those sacrificing them at least occasionally – achieve a higher level of well-being would be needed). Even if the proper method were chosen, it would be naïve to expect that its results would be unequivocal. Their interpretation would, therefore, be difficult. But there seems to be a way out of this quandary. Instead of investigating into how people understand the concept of well-being or examining what psychological research says about the relations between various axiological rankings and the level of well-being, we can make recourse to purely *normative* considerations and ask whether a conception of well-being which implies the absolute priority of prudence over morality is *normatively attractive.* It might seem that the resolution of this problem is a matter of taste but, arguably, this is not so; there appears to be a simple and plausible test of a normative attractiveness of a (normative) conception of well-being, namely: whether we would be ready to defend it publicly, to present it openly to other people, or, more pertinently, to treat it as a central – legitimate and attractive – economic concept. One can safely assume that most of 'us' (including economists) would be rather embarrassed to admit publicly that their conception of well-being allows for pursuit of moral values only if it does not amount to the sacrifice of the prudential ones. In other words, if economics is not a 'dismal science', it cannot assume an exclusive approach to the normative account of well-being. It must go (and, arguably, it does go) beyond it.

But what an inclusive approach to the normative account might look like? It seems that one can distinguish its two main varieties. The first one assumes that:

> The life of an agent S goes well if and only if (a) S's preference set contains a sufficiently large number of prudential preferences (i.e., preferences expressive of his attachment to prudential goods/values); (b) S chooses prudential goods in many decisional situations; (c_1) in the case of conflict between prudential preferences and moral values, the agent makes a reasoned choice between them on a case by case basis (i.e., prior to the decisional situation, he does not give priority to any of them).

The second one, in turn, assumes that:

> The life of an agent S goes well if and only if (a) S's preference set contains a sufficiently large number of prudential preferences (i.e. preferences expressive of his attachment to prudential goods/values); (b) S chooses prudential goods in many decisional situations; (c_2) in the case of conflict between prudential preferences and moral values, the agent always give priority to moral preferences.

Accordingly, the first one treats moral values as important considerations which cannot be discounted while pursuing prudential values, but not as absolute constraints (only as *prima facie* constraints), whereas the second one treats them as 'trumps' which always win over prudential values, that is as absolute constraints which set impassable limits to pursuing prudential values. We do not intend to decide which of these two varieties of the inclusive approach is the open one; we leave it as an open question. But it is our conviction that *any* of them is more plausible than either the non-normative ones or the exclusive approach to the normative account of well-being.

11.4 Concluding Thoughts

We have presented several accounts of well-being, starting from its two non-normative forms (which we criticised above all for their lack of normative dimension), and then passing to its normative accounts: unconstrained prudential and constrained prudential. We have argued that the constrained prudential account is most plausible. One may ask, however, why we have stopped here, that is, why we have not proposed another account, which would locate moral values (rather than prudential ones) in the centre and treat prudential values only as *prima facie* constraints and/or as playing only a subsidiary role in the achievement of well-being. The answer is twofold. First, the general concept of well-being which we endeavoured to explicate or build is subjective making no recourse – unlike the objective one (e.g. Aristotle's) – to any strong metaphysical assumptions about what it means to be a full/perfect human being.[1] Though, clearly, the conception which we defended – the constrained prudential one – contains also two (modest) objective components: (a) it assumes that the overwhelming majority of human beings cannot achieve subjective well-being, that is, feel satisfied with their life, if they do not realize prudential values, but (b) constrained in some way by moral values. The objective component (b) is all the more prominent since our argumentation for its introduction was primarily normative; we did not assume, though we find it plausible, that unjust people cannot be truly happy/satisfied with their life. Second, we have intended to compare only those conceptions of well-being which can be accommodated within economics – with its focus on the individual pursuit of various goals, including 'worldly'/'material' ones (social position, economic success, etc.), the subjective satisfaction with their attainment, and with its rejection of any stronger metaphysical assumptions. By contrast, it is characteristic for the classical (objective) conceptions of well-being (e.g., Aristotle's, and especially that of the Stoics) that they assert that human beings can achieve well-being even in the abject economic conditions. The sublime detachment of this kind of well-being is admirable but is not what economists understand by this term.

[1] See, e.g., Sumner (1998).

References

Bentham, J. 1907. *An introduction to the principles of morals and legislation: Library of economics and liberty*. https://www.econlib.org/library/Bentham/bnthPML.html. Accessed 24 Jan 2020.

Broom, J. 1999. *Ethics out of economics*. Cambridge: Cambridge University Press.

Griffin, J. 1993. Against the taste model. In *Interpersonal comparisons of well-being*, ed. J. Elster and J. Roemer, 45–69. Cambridge: Cambridge University Press.

Harsanyi, J. 1977. Morality and the theory of rational behavior. *Social Research* 44: 623–656.

Hausman, D. 2011. *Preference, value, choice, and welfare*. Cambridge: Cambridge University Press.

Kraut, R. 2009. *What is good, and why. The ethics of well-being*. Cambridge, MA: Harvard University Press.

Pigou, A.C. 1932. *The economics of welfare*. London: Macmillan.

Qizilbash, M. 1996. Capabilities, well-being and human development: A survey. *The Journal of Development Studies* 33: 143–162.

Schick, F. 1986. Dutch bookies and money pumps. *The Journal of Philosophy* 83: 112–119.

Sen, A. 1977. Rational fools: A critique of the behavioral foundations of economic theory. *Philosophy & Public Affairs* 6: 317–344.

———. 1980. Plural utility. *Proceedings of the Aristotelian Society, New Series* 81: 193–215.

———. 1986. Prediction and economic theory. *Proceedings of the Royal Society of London. Series A, Mathematical and Physical Sciences* 407: 3–23.

———. 1988. *On ethics and economics*. Oxford: Blackwell Publishing.

Sumner, L.W. 1998. Is virtue its own reward? *Social Philosophy & Policy* 15 (1): 18–36.

Varian, H. 2010. *Intermediate microeconomics. A modern approach*. New York: W. W. Norton & Company.

Chapter 12
Identity Theories in Economics: A Phenomenological Approach

Ivana Anton Mlinar and Ricardo F. Crespo

Abstract After the seminal 2001 paper written by George Akerlof and Rachel Kranton, the field of identity economics has increasingly developed. This paper presents a new approach to the definition of economic agents' identity, sketching first the conditions required for an appropriate notion regarding the identity of economic agents. Next, it summarizes earlier views outlined by Akerlof and Kranton, Amartya Sen, Miriam Teschl and Alan Kirman, and John B. Davis. Finally, it introduces a phenomenological approach – following E. Husserl's and K. Wojtyla's contributions – combining 'intentionality', 'position-takings' and 'habitualities' as constitutive features of the experience of the acting personal self, to provide a satisfactory identity concept for economic agents.

12.1 Introduction

The 'anomalies' of standard economic theory – the rational choice theory and the expected utility theory – that emerged in experiments conducted over the last 30 years have forced economics to consider imports from non-economic sciences to explain them. Thus, some new research programs, such as behavioural economics, evolutionary economics, neuroeconomics or the capability approach that take elements from other sciences, have been increasingly developed. Identity economics represents an important new approach. John Davis believes that behind the crisis of standard economics' concept of rationality lies its notion of individual identity (Davis 2011, 3). Identity economics captures the idea that personal identity of an individual – an economic agent – is important for explaining individual economic behaviour.

I. Anton Mlinar
Instituto de Filosofía of the Universidad Nacional de Cuyo, Mendoza, Argentina
e-mail: mlinariv@yahoo.es

R. F. Crespo (✉)
Universidad Austral, Universidad Nacional de Cuyo and CONICET, Mendoza, Argentina
e-mail: rcrespo@iae.edu.ar

P. Róna et al. (eds.), *Words, Objects and Events in Economics*, Virtues and Economics 6, https://doi.org/10.1007/978-3-030-52673-3_12

Personal identity greatly influences people's decisions, including their economic decisions. As Davis states, 'Economics and social science – […] – must make use of some conception of the individual to explain economic life' (Davis 2003a, 22). The notion of identity is important to economics in that it provides a necessary theoretical or philosophical framework underlying our descriptions of individual economic agents. We need such a framework because without it the description of the individual economic agent may fail. The agent might still be modelled mathematically, as in standard optimization analysis, but, unless the underlying description can reasonably be said to identify the agent, there is no reason to believe that such an analysis refers to any particular individual. Indeed, we must be able to justifiably say to 'whom' a description applies if we are going to claim a realistic description.

In addition, the literature on identity and economics upholds that agents' descriptions fail to identify real people. Identifying the economic individual poses an issue. Standard economics endorses an atomistic conception of individuals, as Davis (2003a) argues. However, an individual's multiple social commitments shape his/her sense of identity. Therefore, the atomistic individual conception proves inadequate for economics. Moreover, behavioural economics has shown that individuals often make choices that are influenced by context. Yet, while contexts change, individuals remain the same, and individual identity is forged from choices, experiences and circumstances. Thus, it is relevant to know a person's identity to knowing how he/she will act. In fact, Google, Facebook, Amazon use algorithms to detect users' characteristics, identity, concerns, and tastes in order to offer them goods and services accordingly.

Consideration of an individual's social links provides George Akerlof and Elizabeth Kranton the kick-off for identity economics. The *Quarterly Journal of Economics,* in its August 2001 issue, published an article entitled "Economics and Identity" by Akerlof and Kranton. They drew the definition of identity as 'a person's sense of self' (2000, 715) from social psychology.[1] These authors asked how personal identity affects economic facts (2000, 716) and believe that '[i]dentity can account for many phenomena that current economics cannot explain' (2000, 715). They consider identity as 'a new type of externality' (2000, 717).

The notion of identity is also present in Nobel laureate Amartya Sen's 'capability approach'. In addition to his philosophical training, Sen draws from authors like Martha Nussbaum (Nussbaum and Sen 1993) and Michael Sandel (Sen 1999) on identity. Economists-philosophers Alan Kirman and Miriam Teschl have also reflected on identity in economics. Davis, a leading figure in the field of philosophy of economics, has extensively worked on the concepts of identity and economics.

In this paper we will first question what conditions economics requires for a concept of identity. In the second section, we will review some literature on identity and economics, describing the positions adopted by Akerlof & Kranton, Sen, Kirman & Teschl, and Davis. Then, in the third section, we will present an approach that we believe meets the requirements of a notion of identity for economics: a phenomenological perspective on identity. In fact, we will argue that economic reality does not call for a specific notion of human agency and identity, but it needs a notion

[1] On the largely psychological roots of this notion, see Davis (2011, 72–75; 78).

of them in their completeness and unity. Economic agents are not *economic* agents but 'simply' *human* agents, with their whole identity; these agents perform – among other human actions – economic actions, as construed as explained in Sect. 12.1. Finally, a short conclusion will follow.

12.2 A Notion of Economics and Requirements for a Corresponding Concept of Identity

Kirman and Teschl (2004, 62) assert that '[t]he economic agent creates, builds, changes, and learns, is self-reflexive and evaluates her actions'. Crespo (2013, chapter 2) discusses the deep meaning of economic matters – or 'the economy' – from a philosophical standpoint, characterizing economic reality as free, uncertain, and embedded in time. Crespo also notes its subjective character and its social entanglement, exploring three meanings of 'the economic': (1) a metaphoric or improper meaning: human beings are 'economic' insofar as they have needs that they can satisfy using material means; (2) a proper, broad meaning: all decisions and actions geared to the acquisition and use of the goods that satisfy human needs are economic: economic affairs, as they are commonly understood, regardless of their motivations, and (3) a proper, precise meaning: the maximizing character of the use of means in order to achieve ends with those decisions and actions is specifically economic. This last meaning matches the notion adopted by standard economics. Nonetheless, it lacks the richness implied in Kirman and Teschl's description of the actions performed by economic agents and Crespo's characterization of economic reality. We need a theory of agent and identity fitting with descriptions of 'the economic' according to its second meaning – a proper broad meaning. Crespo (2013, chapter 2) also argues that the 'focal' meaning of 'the economic' is economic action.

Specifically speaking about political economy, John Stuart Mill implicitly considers the second and third meanings mentioned above. He first defines political economy as follows:

> What is now commonly understood by the term 'Political Economy' is not the science of speculative politics, but a branch of that science. It does not treat of the whole of man's nature as modified by the social state, nor of the whole conduct of man in society. It is concerned with him solely as a being who desires to possess wealth, and who is capable of judging of the comparative efficacy of means for obtaining that end. (1844/2006, 321)

The last part of the last sentence anticipates the currently prevailing definition of economics – namely, the third meaning of the economic mentioned above: the optimum allocation of scarce means in order to satisfy given ends. However, Mill is aware that this description of political economy involves a simplifying abstraction:

> All these operations, though many of them are really the result of a plurality of motives, are considered by Political Economy as flowing solely from the desire of wealth [...] Not that any political economist was ever so absurd as to suppose that mankind are really thus constituted. (1844/2006, 322)

Therefore, he finally emphasizes the need to consider additional motives for these 'operations' in order to come to a correct explanation and prediction:

> So far as it is known, or may be presumed, that the conduct of mankind in the pursuit of wealth is under the collateral influence of any other of the properties of our nature than the desire of obtaining the greatest quantity of wealth with the least labor and self-denial, the conclusions of Political Economy will so far fail of being applicable to the explanation or prediction of real events, until they are modified by a correct allowance for the degree of influence exercised by the other causes. (1844/2006, 323, see also 326–327)

We provide Mill as an example because, although he propounded the expression 'homo economicus', he recognizes that it is an unreal abstract concept and that real agents' economic decisions can be greatly influenced not only by economic factors but by a plethora of motivations.[2] Adam Smith shares this conception. As Milonakis and Fine (2009, 19) assert, Smith's theoretical edifice is 'rich and multifaceted, encompassing philosophical, psychological, social, historical and economic elements'. Consideration of a plurality of motives for economic actions constitutes a central characteristic of the German Historical School of Economics. Schumpeter (1954, 177–78) remarks that a key feature of this school is that it recognizes that human actions, including economic actions, are not motivated by economic rewards only but are mostly guided by a 'multiplicity of motives', and that it stresses the need to concentrate more on individual correlations than on the general nature of events.

Max Weber (1949, 65–66) distinguished 'specifically economic motives' (almost corresponding to Lionel Robbins' definition of economics) from 'economically conditioned' events and 'economically relevant' activities and situations which are not specifically economic. According to him, we find specifically economic actions only in 'unusual cases' (1978, 15). He considers that there are at least four types of social actions: 'instrumentally rational' (for example specifically economic actions), 'value-rational', 'affectual' and 'traditional', and that it would be 'very unusual' to find actions 'oriented only in one or another of these ways' (1978, 32).

That is, as conceived by classical economic thinkers, the real 'economic action' involves not only the motivations considered by standard economics, but also 'a plurality of motives', 'other causes'. These other motives or causes lie at the root of the characteristics of economic actions as described by Kirman and Teschl and by Crespo: they are rational, psychological, sociological, historical, and ethical – consistent with the second meaning noted above.

Let us use an example to illustrate this point. Buying a new car is an economic action. When buying a car, you make calculations and economic comparisons among car models, taking into account their specific features. However, you may also feel loyalty to a brand, sympathy for the seller, or you may be used to buying cars from a single dealership; you listen to and take into account your wife's tastes and opinions; you might be influenced by the beauty of a specific car, and so on. In short, there are plenty of motivations involved in the transaction.

[2]On the origin of the expression 'economic man' and its meaning according to Mill, see Persky (1995).

Thus, we need a concept of agency and identity that supports all the abilities and characteristics implied in the second meaning above. In fact, these traits do not call for an 'economic' specificity in agents and their identity. Agents performing economic actions are entirely involved in these actions. A notion of 'economic agent' only makes sense for the third meaning the economic; consequently, as the third meaning is an 'idealization', we can only speak about an 'economic agent' as an unrealistic simplification. In the second meaning, there is no economic agent, but 'simply' a human agent, that has to be considered in her completeness. Similarly, we do not need a specific notion of identity for economic actions but 'simply' a human identity – in all its richness. The example of buying a car shows how all kinds of motivations influencing human agency are involved.

Hence, it seems clear that an identity theory becomes necessary to grasp economic affairs because the economic agency is essentially a *human* agency. Indeed, the agent, the *acting-who* must be considered in all her wholeness. In the last section, we will introduce an identity theory that supports the characteristics of human agents when dealing with economic affairs, explaining the link between identity and agency.

The standard economist can be sceptical about the usefulness of considering identity in economics. This is understandable in the context of economics as currently conceived. However, within the conception of economics as a social science, with a methodology that leaves room for prudential reason assessing decisions stemming from an incommensurable plurality of motives, identity happens to be a central motive. This concept differs greatly from today's economics but draws closer to classical political economy. The authors of this paper believe that the 'spirit' of this ancient original thought about economic life should be re-established and that personal and social identity constitute a critical factor for consideration.

12.3 Identity Theories in Economics

Identity theories supported by the following authors vary based on distinct philosophical notions. We will concentrate on Akerlof and Kranton's 'foundational' theory and we will then present other theories as well as some criticisms raised against them by different authors.

12.3.1 Akerlof and Kranton

Using standard economic reasoning tools, such as utility functions and game theory, Akerlof and Kranton show that some presumed anomalies in the standard model (rational choice theory and expected utility theory) can be explained by agents' identity. They state (2002, 1168):

> An individual gains utility when her actions and those of others enhance her self-image. Furthermore, self-image, or identity, is associated with the social environment: People think of themselves and others in terms of different social categories. Examples of social categories include racial and ethnic designations, and in the school context include, for example, "jock" and "nerd." Prescriptions give the ideal, or stereotypical physical attributes and behavior, of people in each category. Individuals then gain or lose utility insofar as they belong to social categories with high or low social status and their attributes and behavior match the ideal of their category.

These authors think that '[w]ithout a model that mirrors this sociology, economic analysis produces only partial answers to key questions' (2002, 1168), arguing that 'identity and norms bring something new to the representation of tastes' (2010, 6). Consequently, they incorporate these new elements to the utility function, viewing the latter as two-fold: the traditional 'standard utility' and 'Identity-utility' (2005, 14; 2010, chapter 3, 17ff). They assert, '[w]e suppose a person chooses actions to maximize her utility, given her identity, the norms and the social categories. She balances her Part 1 standard utility and her Part 2 identity utility' (2010, 18).

This proposal proves positive because it considers new motivations for economic actions, which is a very realistic notion. However, the introduction of these motivations fails because the sociological new inputs lose their meaning in the logic of utility maximizing, which is not their 'natural' logic. The unity of the 'instrumental self' as Elizabeth Anderson (1993, 39) calls the 'self' involved in this economic kind of logic, hinges on the unity of its preferences. It cannot account, she explains, 'for the rational unity of our emotions, attitudes, internalized norms, intentions, and ways of deliberating. In unifying a person's preferences and choices around the achievement of particular consequences, the instrumental view creates discord among other aspects of the self' (1993, 40). The instrumental view only includes other motivations – as Akerlof and Kranton posit – for an instrumental reason – in order to maximize utility – and, thus, 'denaturalizes' these motivations, which do not focus on utility maximization. Instead of subsuming or understanding the instrumental motivation in terms of identity, it subsumes or understands identity in terms of instrumentality. Paraphrasing Anderson (1993, 79), identity 'has global authority' over all possible motivations of actions, while instrumental calculations 'play various local roles within it'. Or, as Martha Nussbaum (1999, 183) puts it, cost benefit analysis only serves as an 'acolyte'.

A consequence of the flip side implicit in Akerlof and Kranton's theory is, as Teschl (2010, 447) remarks, that 'as with all non-market goods, the question is how to evaluate benefits and costs and in the Akerlof and Kranton case, how to evaluate identity gains and losses' (see also Kirman and Teschl 2004, 76–77). Herein lies an incommensurability issue that requires a decision-making process other than a cost-benefit analysis: using practical reason.[3]

[3] See Nussbaum (1999, 182–185), Henry Richardson (1994, 69–86; 209–227). Davis criticizes the inclusion of identity in the utility function, arguing that this is a 'circular explanation': 'the preferences-utility conception of the individual says that if one has one's own (well-ordered) preferences, one can be represented with a utility function and then identified as an independent individual. This, however, only assumes what needs to be shown' (2016a, 24).

We think that the case of Akerlof & Kranton provides a good example of a process described by John Davis (2008b, 365):

> economics, as other sciences, has regularly imported other science contents in the past, and having subsequently 'domesticated' them, remade itself still as economics. In the current situation, for example, behavioral economics — a research program in economics, not in psychology – employs imports from psychology but frames them in terms of economic concerns.

Indeed, Akerlof and Kranton consider psycho-social motivations for economic behaviour, but they 'domesticate' them with the logic of instrumental rationality. Viktor Vanberg (2008, 605–610) reasonably notes that seeking to account for non-economic motives by including them as preferences misses the point. Instrumental motives are outcome oriented. There are non-instrumental motives that are not guided by outcomes but by actions that are valuable by themselves; an instance of 'preferences over actions per se' (Vanberg 2008, 609). This is the case, among others, of actions motivated by people's identities.

In addition, Akerlof and Kranton correctly note that there are inconsistencies between different times in people's lives (2010, 126). Yet, 'what then is the overall identity of the person? It seems that Akerlof and Kranton's approach to introduce identity as motivation for choice leads to the paradox that it dissolves a person's overall identity' (Teschl 2010, 447).

Davis (2011, 81–84) also stresses that Akerlof and Kranton avoid the 'multiple selves' problem by adopting a partial equilibrium analysis: there is no specific criterion to unite people's multiple social identifications. Also, Jason Potts (2008, 4) points out that Akerlof and Kranton offer an 'equilibrium identity' analysis, while he believes that identity is 'by definition a *dynamic disequilibrium,* in which identity is developed and maintained in an entropic open-system context'.[4] Potts argues for a 'generic evolutionary model of identity' in which, 'rather than conceptualizing identity in terms of departures from rationality, identity instead enters economic analysis in terms of the drive to continually recreate and re-invest in individual coherence' (2008, 10).

In short, while Akerlof and Kranton introduce the idea of taking into account identity as a motivation for economic behaviour, they do it in an 'economic-like' way that does not lead to the very identity of economic agents. This is why Davis (2006, 374–377) refers to this position as 'the neoclassical strategy'.[5] Our proposal for a phenomenological approach includes a non-consequentialist notion of identity, i.e., not depending on the outcomes of actions. We will introduce it in Sect. 12.3.

---

[4] Italics in the original text.

[5] See also Ben Fine's critical article, with similar arguments (2009).

12.3.2 Amartya Sen

In *Reason before Identity* (1999), Sen devotes a whole section to the question 'Discovery or Choice?' (1999, 15–19). Sen's view of identity derives from his answer to this question asked by communitarian Michael Sandel. Sandel states that we discover our identities, while Sen believes that we choose them. He softens this view by saying that our choices are not unrestricted (1999, 17) and that sometimes we also make discoveries, but he adds: 'choices have to be made even when discoveries occur' (1999, 19). Davis (2008a) argues that Sen regards having an identity as the most important capability. Given that, for Sen, identity is built by the choices an individual makes, it must be central to the development of individuals' all other capabilities.

Some authors referring to Sen speak about a metaphysical deficit in his view, which hinges on an insufficient conception of human nature. Crocker (1992, 588) asserts that neither Sen nor Nussbaum is trying to ground their ethical proposals on a metaphysics of nature or an account of a trans-historical human essence. Des Gasper (1997, 288ff; 2002, 442, 447, 449–450) complains about Sen's 'thin' conception of the person, adding that Sen's theory also lacks an elaborated theory of the good (2002, 441). Sabina Alkire and Rufus Black (1997) propose to complete Sen's 'deliberately incomplete approach' with John Finnis' practical reason principles. With a more positive approach, Séverine Deneulin (2002) argues that the policies undertaken according to the capability approach (CA) need to be guided by a perfectionist view of the human good. Ananta Giri (2000) regrets the lack of a creative and reflective self in Sen's notions.[6] Benedetta Giovanola (2005) argues for the expansion of Sen's notion on the human person in Marxian terms. The very diverse orientations of these suggestions to overcome Sen's incomplete definitions points to the difficulties of establishing a conception of human beings. However, a minimum notion would help to provide the grounds for a basic guide for social and economic policymaking and would consequently improve CA's operating nature.

12.3.3 Kirman and Teschl

Alan Kirman and Miriam Teschl assert that standard economics has answered the question *what* economic agents are by describing them as maximizing beings who follow their preferences and constraints represented by a utility function. More recently, economics has answered *where* people are locating them in a specific network and society. Akerlof and Kranton's model considers *what* and *where* a person is, but not *who* she is (2004, 73). They propose an answer – new in economics – about *who* she is: 'a self-reflexive human being who has the capacity of actively

[6] Sen's concept of commitment (1977, 2002) seems, however, to entail a reflective self, see Davis (2008a, b).

discovering and consciously creating her identity within a given social context' (2004, 63). These authors believe that this answer is philosophical and goes beyond psychology or social identity. They rely on Derek Parfit's (1984) 'complex view of identity' and on Pierre Livet's (2004, 2006) account of personal identity when presenting their 'who-identity model of identity' (2004; 2006, 303). From Parfit, they draw the idea of the key role of continuity in personal identity (2006, 316). Livet views personal identity as the interplay between two identity functions: personality – more stable – and social status. These ideas inspired Kirman and Teschl (2006, 316–317) notion on who-identity. They describe it as follows:

> In our view, the identity of the economic agent is not characterized by a given and unchanging preference ordering or orderings, but reflects rather a process of continuity and change, i.e. an interplay of three different aspects of a persona that evolve over time: *what* she currently is and does, *who* she wants to be and *where* she chooses to participate, that is, to which social group she chooses to belong. Each of these aspects will correspond to a vector in the characteristics space, a list, undoubtedly very long, of all relevant features of that aspect. The three chosen vectors can be thought of as forming the 'corners' of a *triangle of identity* that moves and changes in a space of characteristics.

Who a person wants to be is her desired self-image, and, to become who she wants to be, the person will choose to identify with the corresponding social groups. However, both the desired self-image and the social groups can change and influence each other.

This proposal accounts for what the person wants to be and, consequently, what she wants to choose and what social group she wants to join, as well as subsequent possible changes. Yet, the question remains as to who the 'who' that underlies all these desires, choices and changes is. In fact, Davis (2011, 197) believes that Kirman and Teschl reproduce Akerlof and Kranton's 'multiple selves' problem – albeit more dynamically. Davis argues, 'How can someone be said to have a personal identity if what that involves continually changes?' Indeed, this is essentially the conclusion of Horst, Teschl, and Kirman (2007, 23), who say that 'personal identity of individuals is relatively weak'.

12.3.4 John Davis

For the past few years, John B. Davis has focused on the analysis of the concept of identity underlying different economic currents and his own concept of identity. Beginning with a book chapter in 2001 and spanning two books (2003b, 2011) and a great number of articles, this topic remains the main focus of his intellectual work to this day. This section will not deal with Davis' appraisal of identity notions in economic theories but with his own concept, especially centring on the ideas expressed in his 2011 book *Individuals and Identity in Economics*.

Davis (2011, 4) calls his approach to identity 'an ontological-criterial approach to identity'. He wonders 'what the *concept* of an individual requires, or what fundamental criteria are involved in referring to things as individuals' (2011, 4). He establishes two criteria: first, the *individuation criterion*, representing individuals as 'distinct and independent beings' (2011, 5). He adds that 'for individuals to be distinct and independent, they must hold together as single whole beings and cannot fragment or break up into multiple selves' (2011, 9). Second, he refers to the *re-identification criterion*, by which those individuals can be re-identified as distinct and independent, despite changes in many individual's characteristics (2011, 5).

Inspired by Sen and Pierre Livet (2006), Davis holds that the individual has a special personal identity capability, 'interpreted as a capability for maintaining and developing an account of oneself in changing interactions with others' (2011, 188). He labels this position 'the capabilities conception of the individual' (2011, 170). Individuals have several changing capabilities. The danger to be avoided is the possibility of transforming the individual in a set of multiple selves, not a unified single being.

He believes this problem can be overcome with 'self-narratives' – 'discursive accounts people keep of themselves' (2011, 183) – that allow people to '*construct* personal identities for themselves in the form of autobiographies' (2011, 171).[7] The identity capability is people's ability to organize themselves through a self-narrative (2011, 190). Self-organizing allows people to have enduring personal identities (2011, 209). Self-narratives are 'evolutionary, open-ended, and generally do not get resolved, because people are continually engaged in developing their capabilities and this continually creates new possibilities for how their narratives will proceed' (2011, 209).

Rather than an individual task, this implies a mutual influence of personal and social identities: 'who they [individuals] are is socially influenced, while at the same time they are a part of the social world because they influence it as well' (2011, 213). Thus, self-narratives are both individual and social.

Miriam Teschl (2011, 79)[8] describes Davis' position on identity:

> This evolution and development of capabilities occurs through social interaction in society. Conflict is important here: different capabilities arise out of different social identities, but it is the conflicts between identities that generate the need to engage in self-organizing processes. Social identity has two aspects for individuals. One is *relational* and concerns an individual's engagement with others from a particular position or role that they occupy using first-person, i.e., self-reflexive, representations. The other is *categorical* and concerns the collective aspect of their identity, assessed from a third-person perspective. Over their lifetimes, individuals keep narrative accounts of themselves, which is a way to reflect on conflicts that their social identities may create, and this engagement and self-examination is what constitutes their personal identities. Indeed, personal identity is an evolving narrative, but it does not necessarily have to be a single, continuous story. It is rather a succession of ongoing conflict-solving discursive accounts, which also help the individual reflect upon the past and project themselves into the future. It is a way of being influenced by and influencing the social structure in which the individual evolves. The individual is thus socially embedded, and yet each self-narrative is highly individualized.

[7] Italics added. From a philosophical point of view the idea of constructing identities has Kantian reminiscences and the idea of constructing them through self-narratives has links with contemporary conceptions of language.

[8] Italics added.

For Davis (2011, 204), the relational social identity proves pivotal for other social identities and unifies them.

More recently, Davis (2016b, 23) has stressed the idea of reflexivity linked with identity: 'individual behavior and identity need to be understood in terms of some sort of capacity to reflexively orient on that behavior and identity, a type of idea which has had little place in the theory of decision-making in economics, with a few exceptions'. He uses Sen's notion on identity and self-scrutiny as an example, associating it with John Searle's idea of preferences as 'the product of practical reasoning' (2001, 253). However, the question remains, who is the individual that self-narrates or reflects? Which is her identity?

In the next section we present an alternative theory which, in our opinion, is consistent with the concept of economic theory described in Sect. 12.1.

12.4 An Alternative Proposal for the Definition of Personal Identity

Kirman and Teschl have insightfully noticed that considering *what* and *where* a person is does not meet all the requirements for an identity-model, but *who* she is actually does (2004, 73). However, we find that a who-identity is not explained either by merely recognizing a continuity factor or by admitting an individuation criterion, as Davis suggests (2011, 4–5); then, both individuation and re-identification criteria fail to characterize *personal* identity: they may apply to any thing or being but only as thing or being – that is, just different from other individuals belonging to the same species. The essentially *personal* condition of identity is missing, and so is identity itself. As shown by the *who*-formula itself, identity starts with a first-person demand, which, of course, may have and indeed does have third-person correlates, as indicated below.

Most contemporary philosophical discussions on personal identity[9] still factor in John Locke's core account of personal identity, in terms of recollection of past experiences or psychological conscious experience of oneself as a means to ensure that continuity factor needed among multiple possible changes over time: 'and as far as this consciousness can be extended backwards to any past Action or Thought, so far reaches the Identity of that Person' (Locke 1975, 27). The fact that a person persists over time does not weigh more heavily than some other facts generally spelled out in either biological or psychological terms, or both. These kinds of interpretation, usually known as complex view, analyse personal identity in terms of simpler relations. This theory not only leads to inconsistences but also fails to explain personal identity.[10] Biological and psychological continuity (in all its possible realizations:

[9] See Shoemaker and Swinburne 1984, Parfit 1984, Williams 1970 and Lewis 1986, among others.

[10] Thomas Reid formulated one such inconsistency by considering the example of a person who can now remember her first day in high school but cannot remember her first day in primary

memory, personality, projects, preferences…) may be regarded as epistemic criteria for an individual diachronic identity, but it provides neither the necessary nor sufficient conditions for *personal* identity.[11] The question of what it takes for a person to persist over time is different from the question of how to find out whether a person at one time is identical to a person at another time. Epistemic criteria for recognizing personal identity over time must not be confused with criteria for identity itself.

Here we present another conception, a phenomenological approach to personal identity, as a contribution to getting a grasp of *the acting personal self* from a first-person perspective. Husserl's critique of Locke's understanding of conscious experience narrows down to remarking that Locke, like many others in the history of philosophy, did not understand intentionality (1956, 76; 92; 110; 112; 114). Husserl states, 'If one has no insight into what is essential to intentionality and into the specific method that belongs to it, one can also not acquire an insight into what is essential to personality and personal accomplishments' (1968, 221).

The phenomenological understanding of the intentionality of consciousness allows us to formulate a theory of personal identity that (1) can account for the continuity of consciousness over time, (2) provides an account of an aspect of what it means to be a person – namely to be able to *appropriate* one's past actions and thoughts as one's own (not merely to remember them or not) – and (3) gives an original answer to the question of personal identity, establishing what the identity of a person over time involves.

In phenomenological terms, intentionality is the basic feature of consciousness – that is, to be always conscious *of something*, to be always geared towards something, to transcend oneself (Husserl 1984, §§ 9–21; 1977, § 36). In other words, intentionality is a relational capacity. However, where a person is concerned, intentionality is not just any relational capacity. Merely experiencing the world and

school, although, on her first day in high school, she could remember her first day in primary school (Reid 2002, 262). Another inconsistency appears when considering identity during sleep: the insistence of Locke himself about the necessarily self-aware nature of our thoughts challenges the possibility that there could be self-aware thought during sleep of which we have no recollection (Locke 1975, II, 1, §10). He even argues that my inability to recollect the thoughts I presumably entertained during sleep leaves open the possibility that they could belong to another person (1975, II, 1, §11).

[11] Knowing everything about bodily and psychological properties and their relations would still leave the question of personal identity unanswered. Consequently, personal identity is conceivable in the absence of psychological and bodily relations. An argument points, for example, to changes of body and psychology (see e.g. Swinburne 1984, 22–3). I can conceive myself as having your body and psychology and you as having mine (more than a thought experiment, it is, by the way, the experience of thought insertion and delusion of control, very common in schizophrenic patients). I could also imagine that I might not have existed, but that instead someone else exists with the same life and body that I actually have. If these scenarios really are metaphysical possibilities, then psychological or bodily relations are neither necessary nor sufficient for personal identity: there is a possible world where I exist without the bodily and psychological properties that I actually have, and another one where the bodily and psychological properties I actually have belong to another person.

others does not, according to Husserl, make us persons yet (animals also do it, which may be viewed as a rudimentary first-person perspective, in contrast to a robust one[12]).

For Husserl (1973, 196), us being persons originates in us performing a specific kind of intentional act, which he calls 'position-taking' (*Stellungnahme*). In the same sense, Wojtyla asserts that "knowledge about man and his world has been identified with the cognitive function [...] And yet, in reality, does man reveal himself in thinking or rather in the actual enacting of his existence? – in observing, interpreting, speculating, or reasoning [...] or in the confrontation itself when he has to take an active stance upon issues requiring vital decisions and having vital consequences and repercussions?" (Wojtyla 1979, vii–viii).

To be a self – namely, to be a person – means, first, not to be a natural object – that is, an 'I' does not appear as a dependent part of causal connections, as a mere individual with its specific essential features, but it emerges in motivational connections of intentional subjects. This can be understood considering that causality is the fixed and empirical legality of physical nature, characterized by certainties in expectations, whereas motivation serves as the basic principle of consciousness – in other words, of the subjective condition. Husserl (2004, 299) asserts, 'It should be taken into account that this causality of nature, [...], is radically different from the causality of motivation that purely reigns inside the sphere of the mental, of the sphere of the immanent subjectivity. In the case of causality of motivation, the necessity of the connection is *comprehensible*'. For him, 'causality in the physical nature is nothing else than a fixed empirical regulation of coexistence and succession, always given in the experience in form of certainties in expectation' (1960, 134).

The intentional character of an 'I' entails taking a stance towards things, towards the world and others, not merely relating in a fixed, predictable way but in a comprehensible manner. It consists of more than perceptual, wakeful awareness, but it does not involve other higher-order activities in its original core sense. Taking a stance does not include making explicative or comparative judgments – higher-order activities. It is simply the defining feature of the personal subjective condition of the self; then, it is the "quality" of the subjective way to view or connect with objects (facts, etc.), revealing their traits but not as imposed features. Subjective receptivity adopts this form: specific availability directions are in the person, influencing the (theoretical, axiological or practical) way she handles an object (facts, etc.).

All position-taking occurs as a specific form of intentional directedness towards being, values, or goals. By means of a progressive position-taking exercise, a stable, "sedimented" character is shaped – that is, the person becomes determined, more receptive to specific dimensions or directions of the availability of things or facts. That is why, as persons, we become characterized by a *habituality* originated in our activity (Husserl 1966, 360). Our past position-takings remain alive to the extent that they shape our future active life. Thus, the person correlates with a horizon of

[12] Cf. Baker (2015, 156).

sense. To relive the past is to remember, but 'habituality' does not mean remembering a past position-taking. Position-takings have their own way of shaping the present, enduring as characteristics of the self, who is the agent responsible for all the positions she has taken.

As the bearer of such enduring position-takings, the ego is always more than the source of its positing, since it is, as a personal ego, also the product of this positing. Through the concept of action, Wojtyla also aims to show not only the person's fundamental experience of being the cause of her own actions, but to be self-determining, that is, – as Buttiglione interprets – the person not only moves beyond the body and the psyche by transcending them but also integrates them in action (1997, 144). The person reveals and realizes herself in action.

As a self with a personal history in the form of enduring interests, choices, and convictions, I am not just aware of a perceptually appearing surrounding; rather, I am aware of this surrounding as displaying my interests, goals, and projects. More precisely, the enduring convictions, projects, and beliefs are expressed in the interest the subject takes in certain cultural, social, scientific, and political practices, which are revealed by the way one experiences the world. What one stands for, moreover, is not some private affair; rather, it shows in the inter-subjective situation via the way we act, think, and talk.

Even if we take the same decisions as others, or if we change our minds constantly, our personal history would, from a phenomenological perspective, still be characterized by an identity: as one always asked to take a position – that is, to relate with a horizon of sense – and as the agent solely responsible for *taking* that position. "In this experience man manifests himself as the person, that is to say, as the highly specific structure of self-governance and self-possession" (Wojtyla 1979, 179). Thus, being a specific person does not mean having a specific corresponding set of features that only belongs to me; then I can indeed share decisions, convictions and traits with others. In fact, there can be one or more individuals with precisely the same set of features as me.

My positions individualize me because they are mine; they originated in me, and, as such, they correlate with an environment. I do not endure like any worldly object, nor do I persist in the way that my habitualities do. The mere ability to recollect a past experience or action does not yet imply the ability to *personally appropriate* it –that is, to take a position, to be intentionally – not causally – related. There is an essential sense of self-governance that denotes the person both as the one who governs herself and as the one who is in a way subjected and subordinate to herself (Wojtyla 1979, 190 ff).

Thus, it may be said that recollection most often reveals continuity, even though the possibility of recollecting is not enough to establish such personal continuity. When I recollect a certain point in my life and still think that I have made the right decision in favour of, for example, a certain profession, the explicit appropriation of my past decision shows that this decision still holds today, both in the sense that I would make the same decision again and that this decision was mine all along, as a habitual and 'sedimented' decision. As such, this decision, even without being re-enacted constantly, has implicitly structured and influenced my other decisions and even prevented me from making other decisions. Indeed, I am still the same person that I was before. Thus, personal identity relies on the capacity to be intentionally related – whether in perception, memory, future or fiction… –, which entails a correlation between myself and a horizon of sense.

As a result, an individual's formal identity appears as the recollecting and recollected subject from a third person perspective: a bundle of characteristics, experiences, features and actions… A genuine personal identity consists of more than the continuity of our conscious awareness and the relative continuity of ourselves over time, changing continuously as we continuously take new stances and revise old ones. A person is hence a whole recognizable by certain habitualities in correlation with an environment. These habitualities provide actual volitional direction, in combination with the motives given in present; then, a person is the subject of free motivations.

Summing up, to be regarded as the only one responsible for taking a stance and not merely as a being that endures over time makes me an ego, separating me from natural objects. Yet, this does not say anything about my personal identity, my uniqueness. My position-taking characterizes me, not only setting up my present way of connecting with the world and others but also establishing my correlation with a horizon of sense. The positions that I take become habitualities, sedimented stances. Even though they can be and actually are revised, some of them (individual and universal) remain unchanged. This fixed relational condition makes up my personal acting self.

Thus, the phenomenological approach to personal identity may fill the gap between hitherto considered formal identity and agency. Formal perspectives of identity fail to grasp that ultimate sense of what is personal and, therefore, fall short of explaining *personal* agency. That intuited "new type of externality" (Akerlof and Kranton 2000, 717) may be found as the expression of the intentional as the motivated way of connecting with the world through habitualities. Davis' individuation and re-identification criteria were considered as grounded on the capability for developing accounts of oneself that give coherence to changes and different stories, emerging like different 'selves' (Davis 2011, 188). However, how could first-personal givenness be brought about by narrative structures? An account of self that disregards the basic structures and features of our intentional-experiential life is not fundamental, and this is the first-person perspective with the primitive form of self-reference that it entails. In order to tell stories about one's own experiences and actions, one must already hold a first-person perspective. Personal identity can be found here, and it may, in turn, correlate with narratives.

Moreover, a certain dimension of inaccessibility and transcendence that characterizes others – the reason why the other is an other – comes precisely from the fact that they are also selves, with their own irreplaceable, unique first-person perspectives.

12.5 Conclusion

In the introductory section we have discussed the importance of identifying the agent in economics. This depends on a particular vision of economics that has been sketchily developed in Sect. 12.1. In Sect. 12.2 we have introduced the concepts of identity in economics proposed by Akerlof and Kranton, Sen, Kirman and Teschl, and John Davis. Finally, in Sect. 12.3, we have looked for another theoretical framework, presenting a phenomenological position about personal identity.

It seems that neither memory nor mere continuity or recollection embody the *who*, the personal self. The mere formal identity of myself as the self that is both recollecting and recollected, or as the individual capable of being represented as a distinct and independent being and that can also be re-identified despite changes – according to Davis – is not a *personal* identity but an individual diachronic identity akin to that of any other object or being. These traits could prove valid epistemic criteria to recognize identity and its continuity over time, but they must not be confused with criteria for identity itself. The key notion is intentionality – actually, intentionality as the essential and structural feature of a personal self: as position-taking. To take a position or stance means to enter into a motivational (not mere causal) level of relationship with the world and others and, therefore, to generate 'habitualities' over time – a 'sedimented' structure informing my actual volitional life in correlation with a horizon of sense. 'Habitualities' are individualized not because of their specific content (that can be revised, changed and even shared with others), but because of their *mineness* – their first person perspective. The notion of a true self as the persistent core of personal identity lies, however, on the ultimate definitiveness of my position-takings: those personal convictions that I experience as an individual calling and values that claim for universality. Thus, decisions, projects and preferences supporting my true self are of a capital importance, and following them leads to habitual attitudes – a sedimented structure that informs my present and future experience of the environment.

If an 'economic action' is a typical human action, all of human agents' motivations and characteristics participate in it. The specificity of economic action – broadly understood – does not call for human agent and human identity specificity; quite the contrary, it requires the consideration of them in their full wholeness and unity.

A phenomenological account considers the person as a whole, and, therefore, the above-mentioned human characteristics of people's actions in relation to economic matters are also involved. People are particularly understood as free, facing uncertainty, but having a specific style when connecting to the world, gained through their position-takings. Hence, they build their own identity in relation to the environment and to others. All the former qualify as characteristics of economic affairs and agents, as described previously. Economic agents are not different from human agents. Every one of them is a singular person who takes positions that make her who she is, becoming *habitualities*, sedimented stances. She decides and acts based on this personal identity, which is obviously richer than any economic maximiser agent.

References

Akerlof, G., and R. Kranton. 2000. Economics and identity. *Quarterly Journal of Economics* CXV: 715–753.

———. 2002. Identity and schooling: Some lessons for the economic of education. *Journal of Economic Literature* XL: 1167–1201.

———. 2005. Identity and the economics of organizations. *Journal of Economic Perspectives* 19: 9–32.

———. 2010. *Identity economics: How our identities shape our work, wages, and well-being*. Princeton: Princeton University Press.

Alkire, S., and R. Black. 1997. A practical reasoning theory of development ethics: Furthering the capabilities approach. *Journal of International Development* 9: 263–279.

Anderson, E. 1993. *Values in ethics and economics*. Cambridge, MA: Harvard University Press.

Baker, L.R. 2015. Selfless persons: Goodness in an impersonal world? *Royal Institute of Philosophy Supplement* 76: 143–159.

Buttiglione, R. 1997. *Karol Wojtyla. The thought of the man who became Pope John Paul II*. Cambridge: William B. Eerdmans Publ. Co.

Crespo, R.F. 2013. *Philosophy of the economy. An Aristotelian approach*. Dordrecht: Springer.

Crocker, D.A. 1992. Functioning and capability: The foundations of Sen's and Nussbaum's development ethic. *Political Theory* 20: 584–612.

Davis, J.B. 2001. Agent identity in economics. In *The economic world view*, ed. U. Mäki, 114–131. Cambridge: Cambridge University Press.

———. 2003a. *The individual in economics. Inaugural lecture. Faculty of Economics and Econometrics*. Amsterdam: Amsterdam University Press.

———. 2003b. *The theory of the individual in economics*. London/New York: Routledge.

———. 2006. Social identity strategies in recent economics. *Journal of Economic Methodology* 13: 371–390.

———. 2008a. The conception of the socially embedded individual. In *The Elgar companion to social economics*, ed. J.B. Davis and W. Dolfsma. Cheltenham: Edward Elgar.

———. 2008b. The turn in recent economics and the return of orthodoxy. *Cambridge Journal of Economics* 32: 349–366.

———. 2011. *Individuals and identity in economics*. Cambridge: Cambridge University Press.

———. 2016a. Economics, neuroeconomics, and the problem of identity. *Schmollers Jahrbuch* 136: 15–32.

———. 2016b. *The sea battle tomorrow: The identity of reflexive economic agents*. Paper presented at the 3rd international conference economic philosophy, Aix-en-Provence, June 15–16.

Deneulin, S. 2002. Perfectionism, paternalism and liberalism in Sen and Nussbaum's capability approach. *Review of Political Economy* 14: 497–518.

Fine, B. 2009. The economics of identity and the identity of economics? *Cambridge Journal of Economics* 33: 175–191.

Gasper, D. 1997. Sen's capability approach and Nussbaum's capabilities ethic. *Journal of International Development* 9: 281–302.

———. 2002. Is Sen's capability approach an adequate basis for considering human development? *Review of Political Economy* 14: 435–461.

Giovanola, B. 2005. Personhood and human richness: Good and well-being in the capability approach and beyond. *Review of Social Economy* LXIII: 249–267.

Giri, A.K. 2000. Rethinking human well-being: A dialogue with Amartya Sen. *Journal of International Development* 12: 1003–1018.

Horst, U., A. Kirman and M. Teschl. 2007. *Changing identity: The emergence of social groups*. GREQAM Groupement de Recherche en Economie Quantitative d'AixMarseille, Document de Travail n 2006-51.

Husserl, E. 1956. *Erste Philosophie (1923/4)*. Erste Teil: *Kritische Ideengeschichte* [*First philosophy (1923/24)*. First part: *The critical history of ideas*], ed. Rudolf Boehm. The Hague: Martinus Nijhoff.

———. 1966. *Analysen zur passiven Synt*hesis. Aus Vorlesungs- und Forschungsmanuskripten, 1918–1926 [*Analyses of passive synthesis. From lectures and research manuscripts, 1918–1926*], ed. Margot Fleischer. The Hague: Martinus Nijhoff.

———. 1968. Phänomenologische Psychologie. Vorlesungen Sommersemester. 1925 [Phenomenological psychology. Lectures from the summer semester. 1925], ed. Walter Biemel. The Hague: Martinus Nijhoff.

———. 1973. *Zur Phänomenologie der Intersubjektivität. Texte aus dem Nachlass*. Zweiter Teil. 1921–28 [*On the phenomenology of intersubjectivity. Texts from the estate*. Second part. 1921–28], ed. Iso Kern. The Hague: Martinus Nijhoff.

———. 1977. *Ideen zu einer reinen Phänomenologie und phänomenologischen Philosophie*. Erstes Buch: *Allgemeine Einführungin die reine Phänomenologie 1*. Halbband: Text der 1–3. Auflage – Nachdruck [*Ideas pertaining to a pure phenomenology and to a phenomenological philosophy*, first book: *General introduction to a pure phenomenology*. First half binding. Text of the 1–3 editions. Reprint], ed. Karl Schuhmann. The Hague: Martinus Nijhoff.

———. 1984. *Logische Untersuchungen*. Zweiter Teil. *Untersuchungen zur Phänomenologie und Theorie der Erkenntnis*. In zwei Bänden. [*Logical investigations*. Second part. *Investigations concerning phenomenology and the theory of knowledge*. In two volumes], ed. Ursula Panzer. The Hague: Martinus Nijhoff.

———. 2004. *Einleitung in die Ethik. Vorlesungen Sommersemester 1920 und 1924* [Introduction to ethics. Lectures of the Sommersemester 1920 and 1924], ed. Henning Peucker. Dordrecht: Kluwer Academic Publishers.

Husserl, E.. Unpublished Manuscripts: A V 21/84b.

Kirman, A., and M. Teschl. 2004. On the emergence of economic identity. *Revue de philosophie économique* 9: 59–86.

———. 2006. Searching for identity in the capability space. *Journal of Economic Methodology* 13: 299–325.

Lewis, D. 1986. *On the plurality of worlds*. Oxford: Blackwell.

Livet, P. 2004. La pluralité cohérente des notions de l'identité personelle. *Revue de Philosophie Economique* 9: 29–57.

———. 2006. Identities, capabilities and revisions. *Journal of Economic Methodology* 13: 327–348.

Locke, J. 1975. *An essay concerning human understanding*. ed. Peter H. Nidditch. Oxford: Clarendon Press. References are made to the Book, Chapter, and Section e.g. (II, 26, §13).

Mill, J.S. 1844/2006. *Essays on some unsettled questions of political economy* (Essay V: 'On the definition of political economy; and on the method of investigation proper to it'). In *Collected works of John Stuart Mill*, vol. 4. Indianapolis: Liberty Fund.

Milonakis, D., and B. Fine. 2009. *From political economy to economics*. Abingdon: Routledge.

Nussbaum, M.C. 1999. Virtue ethics: A misleading category? *The Journal of Ethics* 3: 163–201.

Nussbaum, M.C., and A. Sen, eds. 1993. *The quality of life*. Oxford: Clarendon Press.

Parfit, D. 1984. *Reasons and persons*. Oxford: Oxford University Press.

Persky, J. 1995. Retrospectives. The ethology of *Homo Economicus*. *Journal of Economic Perspectives* 9 (2): 221–231.

Potts, J. 2008. Economic evolution, identity dynamics and cultural science. *Cultural Science Journal* 1 (2). http://doi.org/10.5334/csci.15; https://culturalscience.org/articles/10.5334/csci.15/.

Reid, Th. 2002. *Essays on the intellectual powers of man*, ed. Derek R. Brookes. University Park: Pennsylvania State University Press.

Richardson, H.S. 1994. *Practical reasoning about final ends*. Cambridge: Cambridge University Press.

Schumpeter, J.A. 1954. *Economic doctrine and method. An historical sketch*. New York: Oxford University Press.

Sen, A. 1977. Rational fools: A critique of the behavioral foundations of economic. *Philosophy & Public Affairs* 6: 317–344.

———. 1999. *Reason before identity*. New Delhi: Oxford University Press.

———. 2002. *Rationality and freedom*. Cambridge, MA: The Belknap Press of Harvard University Press.

Shoemaker, S., and R. Swinburne. 1984. *Personal identity*. Oxford: Blackwell.

Teschl, M. 2010. Identity economics: Towards a more realistic economic agent? Review of *Identity economics: How our identities shape our work, wages, and well-being*. *Journal of Economic Methodology* 17: 445–452.

———. 2011. Review of John B. Davis's *Individuals and identity in economics*. *Erasmus Journal for Philosophy and Economics* 4: 74–82. http://ejpe.org/pdf/4-2-br-2.pdf.

Vanberg, V. 2008. On the economics of moral preferences. *American Journal of Economics and Sociology* 67 (4): 606–628.

Weber, M. 1949. *The methodology of the social sciences*. New York: The Free Press.

———. 1978. *Economy and society*. Berkeley: University of California Press.

Williams, B. 1970. The self and the future. *Philosophical Review* 79: 161–180.

Wojtyla, K. 1979. *The acting person. Analecta Husserliana, the yearbook of phenomenological research*. Vol. X. Dordrecht: D. Reidel Publ. Co.

Chapter 13
Temporal Structures of Justification in the Economic Analysis of Law: Legal Philosophy and Free Will

Kevin Jackson

Abstract In an appropriation of jurisprudence away from moral philosophy in the name of advancing positive science, the economic analysis of law repudiates judicial responsibility in regard to human freedom and rights. I argue that the moral freedom and responsibility presupposed by a system of legal rights should not be eclipsed by narrow causal determinism but instead upheld by proper judicial discernment of wider temporally balanced structures of legal rights within such a system. Indeed, this is part of a wider claim that moral philosophy at large proceeds at its best by means of a temporally holistic reasoning process. I am interested in exposing and challenging theoretical bias and unacknowledged assumptions about time-value – in particular the privileging of future outcomes – lurking behind the economic analysis of law philosophy.

13.1 Economic Analysis of Law

Economic analysis of law uses concepts from micro-economic theory to analyse legal rules and institutions. Such an approach adopts the idea of rational action that is at the heart of micro-economic theory.

The notion of rational action in economic theory holds that each agent will act to maximize her or his preferences no matter what environment is being considered. Preferences are rankings of various elements in one's domain of preference. An agent's domain of preference is made up of the things that fundamentally matter to him or her. For typical consumer behaviour models, an agent has fundamental preferences over consumption bundles. The consumer's decision problem consists of

---

This paper was given at Blackfriars Hall, Oxford University on Oct 4, 2019 at a related conference.

K. Jackson (✉)
Fordham University, New York, NY, USA
e-mail: kjackson@fordham.edu

P. Róna et al. (eds.), *Words, Objects and Events in Economics*, Virtues and Economics 6, https://doi.org/10.1007/978-3-030-52673-3_13

choosing a consumption bundle from the set of feasible consumption bundles (feasibility being defined by prices of goods and the income of the agent).

Normally the scenario for an economic analysis of law is slightly more complicated. An agent will have preferences concerning a set of consequences, like income, health or wealth. And the agent chooses an action that partly determines which consequence is realized. Ordinarily the domain of preference differs from the domain of choice. That is usually because an agent chooses a strategy that, coupled with the strategy choices of other agents, together determines the consequence at hand.

Richard Posner advances two theses that are characteristic of the economic analysis of law (Posner 2014). What is termed the "positive" thesis says that, as a factual matter, the legal rules of common law are efficient. His second thesis, usually referred to as the "normative" thesis, says that the legal rules of common law ought to be efficient.[1] The concept of efficiency in this context means maximization of social willingness-to-pay. Other law-and-economics scholars sometimes employ "efficiency" in the Pareto sense.

Under the guise of the economic analysis of law approach, judges become emboldened to exercise activist inclinations, using legal rhetoric about rights *instrumentally* to advance various social agendas. For the law and economics movement, such agendas are grounded in a peculiar ideology of rational choice, efficiency, pareto optimality, and wealth maximization. Propositions about rights are taken merely on an *as-if* basis – it is as if we had rights, but they don't really exist. All that exists are competing preferences to be understood and arranged according to causally determined scientific laws.

13.2 Theoretical Time-Privileging

This paper posits that the economic analysis of law fails to give any satisfactory normative justification for its peculiar *theoretical privileging of the future over the past*. Judicial decision making, according to the economic analysis of law mindset, is portrayed as giving justifications exclusively in terms of future outcomes.[2] The

[1] Throughout Posner's work stress is placed on the expected efficient results normally accomplished through free trade and voluntary exchange. In Posner's law-and-economics world the good is prior to the right. Yet it seems that the normative appeal of voluntary exchange and the legal rights that uphold this institution should not rest upon the resultant wealth maximization and efficiency that are anticipated to ensue. Rather, what ought to be treasured most is moral respect for free will and chosen allegiance to rule of law – the system of legal rights and responsibilities as such – rather than the economic outcomes they enable. If market solutions are deemed superior it is because they are the intended outcome of free exchanges (O'Driscoll 1980).

[2] In this respect, the economic analysis of law falls within the criticism Ronald Dworkin levelled against what he termed 'legal pragmatism.' "The pragmatist takes a skeptical attitude toward the assumption we are assuming is embodied in the concept of law: he denies that past political decisions in themselves provide any justification for either using or withholding the state's coercive power. He finds the necessary justification for coercion in the justice or efficiency or some other contemporary virtue of the coercive decision itself, as and when it is made by judges, and he adds

economic analysis of law proceeds squarely in line with the dictates of modern economic theory's obsession with calculations of preferences. By contrast, coming from the opposite direction, the judicial philosophy of 'originalism' harbours its own theoretical bias of time-value, which proceeds from a privileging of the past concerning questions of constitutional intent.[3]

Economic theory regards preferences as a type of empirical data, and it seeks to bypass any regard for the moral content of preferences or for the propensities of preferences to change in the face of moral considerations. However, what economists term 'preferences,' sometimes include peoples' judgments on moral issues. Such judgments are made as a result of moral reflection and argumentation. As such, economic theory also ignores the various temporal value structures that are constitutive of preferences. Such temporal value structures implicate the free will of people making choices in time that are not causally determined nor predicable in advance of the time they are made. The moral decisions made engage further choices about alternative time-values – whether consequentialist (future oriented considerations) or nonconsequentialist (blending past and future considerations) – which together is what enables them to be responsible moral agents. In their reductive treatments of these choices, economists are ignoring the fact that sometimes they are determined by moral judgments arrived at through antecedent moral arguments. So a more complete treatment would take into account the soundness of the moral argumentation that is presupposed by preferences. Accordingly, the best foundation for philosophy of law is not economic analysis and economic theory but rather moral philosophy.

13.3 Temporal Structures of Moral Decision Making

Temporal structures are implicated in moral decision making and in the acts of free-will that follow from those decisions. Consider that at time t, moral agent M, faces choices x or y (x: give money to a homeless person on the street, or y: walk on by). M can freely decide to do x or y. At time t_1, after considering various arguments for and against, M decides to do y.

One significant point is that M's decision to do y is not causally determined in advance of $t_{1.}$ On the contrary, the choice is free, it cannot be predicted in advance, either by M or by an external observer. Like M, our conscious moral decisions are independently significant in their own right. The logic of ethical justification is

that consistency with any past legislative or judicial decision does not in principle contribute to the justice or virtue of any present one. If judges are guided by this advice, he believes, then unless they make great mistakes, the coercion they direct will make the community's future brighter, liberated from the dead hand of the past and the fetish of consistency for its own sake. Of course judges will disagree about which rule, laid down in which circumstance, would in fact be best for the future without concern for the past." Dworkin (1986, 151)

[3] For an account of originalism, see Whittington (2013).

different from the logic of scientific explanation. The importance of making a moral case supporting a decision, and the weight that such a justification carries, is not contingent on, nor even particularly relevant to, any remote causal explanation that scientific study may offer.

A second important point is that M's moral reasoning involves free (non-causally determined) choices about time orientation, horizon, and so on that express and depend upon a variety of M's values. These engagements in time-value factors are not quantitatively measurable clock-time events but rather qualitative touchpoints that figure into moral reflection. M considers, say, the significance of the recently-past fact that he already has given money to charities just this week; M thinks ahead about whether, if he gives the beggar some money he will still have enough cash to cover the subway ride M's about to take; M reflects that often in the past when he has given handouts to people they've used the money for drugs or booze; he recalls a time long ago when he himself was down and out; he wonders what if this was his son or daughter. Further, M makes a rough assessment of whether the beggar seems to be sincere and desperate, based on a complex mix of other past, present and future factors. M thinks about how these kinds of choices and judgments connect to the kind of person M has been, is, and will become; and perhaps M extends a similar process of thinking to the perceived character of the beggar as well. Or M may not give a thought to any of it at all – who knows? At any rate, even M's deciding not to think about it is itself a moral choice.

A third point is that it is meaningful, and in line with what ordinary people do, to either blame or praise M (for his ungenerousness or sound judgment, respectively) if, say, he decides to do y. But ascribing blame or praise is only meaningful on the assumption that M's decision comes from his free will. Determinism holds that your sense that you decided to walk on by comes from exercising your "free will" is delusional – your decision was caused by forces or events from the past. But determinism does not show why any of this is true, nor give convincing reasons why we should suppose that we are living under a grand illusion about our free will. Determinism depicts the decision-making process the way we understand our internal organs to function. You don't will or choose to make your heart beat or intestines digest. In this way determinism misrepresents moral decision making as a cause-and-effect process with no room left for the temporal nuances attending the free moral choices we have identified as figuring into one's ordinary decision-making process.[4]

A fourth point – moving from M's scenario to the context of judicial reasoning – is that in a relevantly similar way to M's moral decision-making, the structure of legal decision-making presupposes and engages a vast and intricate web of time-values, and as such is not necessarily or exclusively portrayable as a project of justifying results solely in terms of future outcomes (efficiency, pareto optimality, wealth maximization). For instance, in civil cases, the well-entrenched legal

[4]According to the research studies of Zimbardo and Boyd (2009), people are psychologically determined to dwell, in quantifiably measurable allocations, on past, present, or future orientations with negative or positive attitudes.

time-value structure of *stare decisis* requires that judicial decisions respect relevantly similar case precedents laid down in the past by the same or a superior court. It is owing to the force of *stare decisis* that a party's legal rights are, in an important sense, pre-existing and thus interpretations of those rights are not hostage to the sort of future-oriented consequentialist justifications advanced by the economic analysis of law.

13.4 Time and Free Will

Henri Bergson sought to mount a challenge against arguments that deny the existence of free will. In *Time and Free Will,* he claims that such arguments stem from confused concepts about time. The concept of time enlisted by physics and mathematics is that time is a measurable construct analogous to dimensions of space. Yet Bergson insists that from the standpoint of human experience, one's life is comprehended as a non-measurable, continuous flow. As such it is different from a succession of distinct, quantitatively measurable states of consciousness. The perceived flow is essentially qualitative, and not subject to quantitative reduction. Insofar as the human personality is expressed and revealed in actions that are not completely predictable, Bergson asserts that human free will is an observable fact. So *Time and Free Will* posits the idea of duration – lived time – in contradistinction from Bergson claims is a spatial understanding of time whereby time is measured and quantified scientifically with the instrument of a clock.

For Bergson the awareness we have of an inner self shows that psychological facts remain qualitatively distinct from other kinds of facts. Indeed, Bergson alleges that, in their pursuit of quantifying and calculating relations within the phenomena of this inner dimension psychologists ('psychophysicists') are implicated in a practice of "an inaccurate psychology, misled by language," (1910, 165) that will "fail to translate completely what our soul experiences" (1910, 164). He takes issue in particular with Fechner's reductive quantitative portrayal of a purported logarithmic lawlike relationship between the intensity of stimuli and the magnitude of corresponding sensations they produce (1910, 62)

$$S = C\int_{o}^{E} \frac{dE}{f(E)}$$

> [B]y invading the series of our psychic states, by introducing space into our perception of duration [the confusion of quality with quantity] corrupts at its very source our feeling of outer and inner change, of movement, and of freedom. (1910, 74)

For Bergson the key is to clear away conceptual confusions in metaphysics and psychology that confound the problem of free will:

> What I attempt to prove is that all discussion between the determinists and their opponents implies a previous confusion of duration with extensity, of succession with simultaneity, of

> quality with quantity: this confusion once dispelled, we may perhaps witness the disappearance of the objections raised against free will, of the definitions given of it, and, in a certain sense, of the problem of free will itself. (xxiii–xxiv)

More recently, Thomas Nagel posits a distinction between two realms of truth concerning ourselves and the place we occupy in the world. One realm concerns our subjective, personal point of view. The other realm concerns the objective, impersonal perspective from which one seeks to understand oneself as belonging to the natural world. Nagel (1986, 115) thinks the free will problem stems from the shift from one realm to the other:

> The objective view seems to wipe out such autonomy because it admits only one kind of explanation of why something happened – causal explanation – and equates its absence with the absence of any explanation at all... [T]he basic idea which it finds congenial is that the explanation of an occurrence must show how that occurrence, or a range of possibilities within which it falls, was necessitated by prior conditions and events.

But is the impersonal point of view Nagel characterizes suitable for handling moral issues concerning responsibility, where such issues surpass scientific questions about freedom? Some scholars have developed views that treat issues of moral freedom as reducible to matters of control that is uncaused. Yet if it is true, as determinism holds it to be, that all our behaviour is determined by external forces, then one is never in control of one's behaviour in the sense that one can be held responsible for it. A scientifically reductive determinism links moral judgments about responsibility to scientific explanations that are rendered on the basis of causation. Free will, from such a standpoint, is an illusion. Common attitudes about praise and blame and punishment, according to views espoused by J.C.C. Smart in his 1961 *Mind* article, need to be adjusted accordingly. The implications of scientific determinism for the philosophy of law are as astounding as they are implausible. Some legal scholars say that because science shows there is no free will, it is always wrong to punish. Instead we should treat criminals medically or reprogram them to change their behaviour. A similar view is put forward by an SEC attorney who couples his advocacy of scientific determinism and denial of moral free will with a thesis about the unreality of time (Gulack 2012).

13.5 Free Will and Legal Rights

Having a right to something means that you have control over others' free will in regard to it. Otherwise, they can do as they please.

Someone violates your right by acting contrary to your free will in regard to your right's object.

The will theory of rights, also known as the "choice theory," allows rights-holders the freedom of choice to insist upon respecting their rights or permitting them to be waived.

Example 1: the right to ownership of my land includes the freedom to do with it as I wish (subject to a set of legal constraints). It is wrong (a rights violation) for someone to interfere with my freedom unless they have a right to so interfere. If someone uses my land without having a right to do so, then I am free to either allow it, or to choose to prevent it by claiming protection of my right from relevant legal authorities (injunction, criminal arrest, tort action for trespass).

Example 2: you cannot be subjected to harvesting of your bodily organs simply because transplants in others could be shown to make society better off. A tort right protects your individual interest in bodily security from any such coerced social redistributions. To lend protection, the tort right prioritizes your right-holder's security interest over conflicting liberty or economic interests of correlative duty-bearer(s). Your tort-based right, in turn, is grounded in an even more fundamental moral right of dignity.

A legal right has a *temporal justificatory structure* that extends from the present both ahead towards the future and backward towards the past. Any theory of law that fails to recognize this complex, multi-temporal nature of rights will be descriptively deficient and normatively implausible.

That is to say, a theory of law should be *temporally holistic* in the way it accounts for the interpretation of rights.

13.6 Holism of Time-Value

Legal rights refer (either in tacit or explicit fashion) retrospectively to values from the past that ought to be respected. Interpretations of rights must bear consistency with settled precedents and statutory authority laid down in the past. This idea is enshrined in the common law doctrine of *stare decisis*. Legal rights also refer prospectively to justifications based on moral objectives. Interpretations of rights must make sense in terms of political morality. Of course, legal rights refer to and exist in the present as well, as political rights that may be enforced on demand via coercive legal structures and adjudicative institutions. The call for their present recognition and enforcement does not require future supplemental legislation or additional law-creating action.

One example (there are others) of a temporally holistic conception of law is found in the work of Ronald Dworkin. Within Dworkin's philosophy of law, there are two fundamental temporal structures – I will call them "theoretical time-value orientations" – that figure into interpretations of law. The first component, which he terms 'fit,' is *retrospective*. The second component, called 'appeal' or 'justification,' is *prospective*. In judicial decision making, a challenge is to develop a theory of law that simultaneously fits with past legal decisions while making the law the best it can be. Judges are bound to seek out historical legal principles that are woven into previously established 'institutional material' and then look forward to improving and further purifying law for the future by rendering the most coherent interpretation of it.

In opposition to the claim that pre-existing legal sources might be contradictory or irreconcilable, Dworkin gives an interpretation according to what he terms the 'unity of value' thesis. The project of interpretation he envisions reconciles values. The reconciliation seeks to show why moral conflict may call for a deeper structure of collaboration to resolve apparent conflicts and sometimes to illuminate a new point of comparison or contrast.

Dworkin presents a version of the drowning swimmer scenario:

> One person clings to a life preserver in a storm that has wrecked her boat; sharks circle her. Two other passengers cling to another life preserver a hundred yards away; sharks circle them as well. You have a boat on shore. You can reach one life preserver in time, but then not the other one. Assuming all three are strangers, do you have a duty to save the two swimmers and let the lone swimmer die? (2011, 280)

> But if we approach the decision in another way – by concentrating not on consequences but on rights – it is far from plain that we should automatically save the greater number. We might think that each victim has an equal antecedent right to be saved, and we might therefore be tempted by a lottery in which each shipwreck victim has at least one-third chance to be saved. (The sharks agree to circle while the lottery is conducted.) (2011, 281)

Whereas most people may, owing to the fact that there are *two* swimmers holding on one of the life preservers, yet only one swimmer clinging to the other, be automatically inclined to save the two. That response follows from a utilitarian (consequentialist) calculation to advance the greatest good for the greatest number. It is not clear, however, that this is the right solution from a principles-and-rights conception. Yet it is not obvious that each of the swimmers that one can aid has an "equal antecedent right to be saved." If it is too dangerous and you would unreasonably risk your own life, then a right to be saved cannot be invoked to force you to give aid. The swimmers can, nevertheless, claim a right to equal concern and respect.

That right to equal concern and respect, in turn, is interpreted in light of a broad moral value: that life has value. So one needs to interpret the available options – lottery to decide among them, or whether the lone swimmer is your spouse, or a gifted violinist or outstanding surgeon – in line with respecting the value of human life (2011, 281).

I wish to drill down further and ask: what does it mean to speak, in such contexts, of having an "*antecedent right*" in temporal terms? Is it a non-temporal antecedence – that is, only a logical sense of antecedent? (As in 'If x, then y' – x is the logical antecedent to y). The idea of antecedent *legal* rights does have a temporal sense, since the rights are *pre-existing* (they are not invented on-the-spot, but rather interpreted), in that they exist already (historically) in settled law. To be sure, in the swimmer case, we are not dealing with legal (institutional) rights but rather with a moral rescue (aid) context. But that context is nevertheless used to illustrate the structure of legal rights and duties. It easily could be a legal case in some places, considering that there is a legal duty to rescue in some countries.

Consider some of the ways that various temporal factors – engaging a host of value judgments – might come into play, particularly in connection with interpreting the "equal objective importance of all human lives":

- One can justifiably choose to save the lone swimmer if, say, it is one's spouse. In that case, there is a special justification that refers to the *past* (the historical act of marriage and resultant relationship). And the justification presumably draws support as well from one's *present* and *future* commitment to the spouse.
- What if the lone swimmer, a convicted child molester, nevertheless came out to save you from drowning last week when you'd fallen overboard? Does his criminal *past* matter *now* and for the *future* (should you forgive and forget)? Is there a present duty to reciprocate in saving his life now in respect of his past saving of your life?

What I would like to draw attention to in such examples (and others) is this: neither the past nor the future dominates in this approach. There is instead a need for intricate temporal balancing that defies any rigid privileging in any one direction or for any given retrospective or prospective distance.

13.7 So What?

Why does this kind of critique of the distorted temporality deployed in the economic analysis of law matter? There is a resurgence in interest in the economic analysis of law, fuelled in no small way by trends toward using digital technologies (including artificial intelligence) for resolving legal conflicts. The empirical-scientific conception of law, which sees legal conflicts as best resolvable through future-oriented assessments of quantifiable data according to objective scales of efficiency, Pareto optimality, welfare maximization of revealed preferences, and so on, lends apparent philosophical support to the agenda of such a research project. This kind of reductive scientism (not scientific method *per se*) should be opposed, not only because of the perils of the ideology of scientism as such, but also because of the theoretical shortcomings of the economic analysis of law exposed herein.

References

Bergson, H. 1910. *Time and Free Will*. Trans. F.L. Pogson. Humanities Press.
Dworkin, R. 1986. *Law's empire*. Cambridge: Harvard University Press.
———. 2011. *Justice for hedgehogs*. Cambridge: Harvard University Press.
Gulack, R. 2012. https://ethicalfocus.org/the-land-of-now-or-how-time-goes-by/. Accessed 13 Sept 2019.
Nagel, T. 1986. *The view from nowhere*. Oxford: Oxford University Press.

O'Driscoll, G.P. 1980. Justice, efficiency, and the economic analysis of law: A comment on fried. *The Journal of Legal Studies* 9 (2): 355–366.
Posner, R.A. 2014. *Economic analysis of law*. 9th ed. New York: Wolters Kluwer.
Smart, J.C.C. 1961. Free will, praise and blame. *Mind* 70 (279): 291–306.
Whittington, K.E. 2013. Originalism: A critical introduction. *Fordham Law Review* 82 (2): 375–409.
Zimbardo, P., and J. Boyd. 2009. *The time paradox: The new psychology of time that will change your life*. New York: Free Press.

Index

P. Róna et al. (eds.), *Words, Objects and Events in Economics*, Virtues and Economics 6, https://doi.org/10.1007/978-3-030-52673-3

The manufacturer's authorised representative in the EU is Springer Nature Customer Service Centre GmbH, Europaplatz 3, 69115 Heidelberg, Germany. If you have any concerns regarding our products, please contact ProductSafety@springernature.com

Printed and bound by CPI Group (UK) Ltd, Croydon, CR0 4YY
30/09/2026
02225696-0012